I0488213

The Source, Discharge, and Chemical Characteristics of Water from Agua Caliente Spring, Palm Springs, California

Edited By Peter Martin

With contributions by Justin Brandt, Rufus D. Catchings, Allen H. Christensen, Alan L. Flint, Gini Gandhok, Mark R. Goldman, Keith J. Halford, Victoria E. Langenheim, Peter Martin, Michael J. Rymer, Roy A. Schroeder, Gregory A. Smith, and Michelle Sneed

Prepared in cooperation with the Agua Caliente Band of Cahuilla Indians

Scientific Investigations Report 2011–5156

U.S. Department of the Interior
U.S. Geological Survey

U.S. Department of the Interior
KEN SALAZAR, Secretary

U.S. Geological Survey
Marcia K. McNutt, Director

U.S. Geological Survey, Reston, Virginia: 2011

For more information on the USGS—the Federal source for science about the Earth, its natural and living resources, natural hazards, and the environment, visit http://www.usgs.gov or call 1–888–ASK–USGS.

For an overview of USGS information products, including maps, imagery, and publications, visit http://www.usgs.gov/pubprod

To order this and other USGS information products, visit http://store.usgs.gov

Suggested citation:
Martin, Peter, ed., with contributions by Brandt, Justin, Catchings, R.D., Christensen, A.H., Flint, A.L., Gandhok, Gini, Goldman, M.R., Halford, K.J., Langenheim, V.E., Martin, Peter, Rymer, M.J., Schroeder, R.A., Smith, G.A., and Sneed, Michelle, 2011, The source, discharge, and chemical characteristics of water from Agua Caliente Spring, Palm Springs, California: U.S. Geological Survey Scientific Investigations Report 2011–5156, 106 p.

Acknowledgements

This study was funded and developed in cooperation with the Agua Caliente Band of the Cahuilla Indians. The authors thank the Agua Caliente Band of the Cahuilla Indians, the Desert Water Agency, and Earth Systems Southwest who provided data and allowed access to their property for sampling springs and wells.

The authors also thank the many U.S. Geological Survey personnel that helped in the completion of this report, including: John Izbicki, Justin Kulongoski, Andy Morita, Laurel Rogers, and Larry Schneider. The authors are especially indebted to Greg Lines (posthumous) who helped develop the original proposal for this study and provided mentorship during the early phases of the study.

Contents

Contents—Continued

Figures

Figures—Continued

Tables

Conversion Factors, Datums, and Abbreviations

Inch/Pound to SI

Multiply	By	To obtain
Length		
inch (in.)	2.54	centimeter (cm)
inch (in.)	25.4	millimeter (mm)
foot (ft)	0.3048	meter (m)
mile (mi)	1.609	kilometer (km)
Area		
acre	4,047	square meter (m^2)
acre	0.4047	hectare (ha)
acre	0.4047	square hectometer (hm^2)
acre	0.004047	square kilometer (km^2)
square mile (mi^2)	259.0	hectare (ha)
square mile (mi^2)	2.590	square kilometer (km^2)
Volume		
gallon (gal)	3.785	liter (L)
gallon (gal)	0.003785	cubic meter (m^3)
gallon (gal)	3.785	cubic decimeter (dm^3)
million gallons (Mgal)	3,785	cubic meter (m^3)
cubic foot (ft^3)	28.32	cubic decimeter (dm^3)
cubic foot (ft^3)	0.02832	cubic meter (m^3)
acre-foot (acre-ft)	1,233	cubic meter (m^3)
acre-foot (acre-ft)	0.001233	cubic hectometer (hm^3)
Flow rate		
foot per second (ft/s)	0.3048	meter per second (m/s)
foot per minute (ft/min)	0.3048	meter per minute (m/min)
foot per day (ft/d)	0.3048	meter per day (m/d)
cubic foot per minute (ft^3/min)	0.0283	cubic meter per minute (m^3/s)
gallon per minute (gal/min)	0.06309	liter per second (L/s)
Acceleration		
milligal (mGal)	0.0001	meter per second squared (m/s^2)
Pressure		
atmosphere, standard (atm)	101.3	kilopascal (kPa)
bar	100	kilopascal (kPa)
Radioactivity		
picocurie per liter (pCi/L)	0.037	becquerel per liter (Bq/L)
Hydraulic conductivity		
foot per day (ft/d)	0.3048	meter per day (m/d)

Conversion Factors, Datums, and Abbreviatons—Continued

SI to Inch/Pound

Multiply	By	To obtain
Length		
centimeter (cm)	0.3937	inch (in.)
millimeter (mm)	0.03937	inch (in.)
meter (m)	3.281	foot (ft)
kilometer (km)	0.6214	mile (mi)
kilometer (km)	0.5400	mile, nautical (nmi)
meter (m)	1.094	yard (yd)
Area		
square kilometer (km^2)	247.1	acre
square kilometer (km^2)	0.3861	square mile (mi^2)
Volume		
cubic meter (m^3)	6.290	barrel (petroleum, 1 barrel = 42 gal)
liter (L)	33.82	ounce, fluid (fl. oz)
liter (L)	2.113	pint (pt)
liter (L)	1.057	quart (qt)
liter (L)	0.2642	gallon (gal)
cubic meter (m^3)	264.2	gallon (gal)
cubic meter (m^3)	0.0002642	million gallons (Mgal)
cubic centimeter (cm^3)	0.06102	cubic inch (in^3)
liter (L)	61.02	cubic inch (in^3)
cubic meter (m^3)	35.31	cubic foot (ft^3)
cubic meter (m^3)	1.308	cubic yard (yd^3)
cubic meter (m^3)	0.0008107	acre-foot (acre-ft)
Flow rate		
meter per second (m/s)	3.281	foot per second (ft/s)
Mass		
kilogram (kg)	2.205	pound avoirdupois (lb)

Temperature in degrees Celsius (°C) may be converted to degrees Fahrenheit (°F) as follows:

$$°F=(1.8×°C)+32$$

Temperature in degrees Fahrenheit (°F) may be converted to degrees Celsius (°C) as follows:

$$°C=(°F-32)/1.8$$

Vertical coordinate information is referenced to the National Geodetic Vertical Datum of 1929 (NGVD of 1929) and North American Vertical Datum of 1988 (NAVD 88).

Horizontal coordinate information is referenced to the North American Datum of 1983 (NAD 83).

Altitude, as used in this report, refers to distance above the vertical datum.

Specific conductance is given in microsiemens per centimeter at 25 degrees Celsius (µS/cm at 25°C).

Concentrations of chemical constituents in water are given either in milligrams per liter (mg/L) or micrograms per liter (µg/L).

Acronyms

AGC	automatic gain control
AWD	accelerated weight drop
bls	below land surface
cc STP/g	cubic centimeters of gas at standard temperature and pressure of 25°C and 1 bar per gram of water.
CDP	Common Depth Point
CFC	Chlorofluorocarbon
DIC	dissolved inorganic carbon
DSC	dissolved-solids concentration
ESA	European Space Agency
GIDS	gradient and inverse distance squared
GPS	Global Positioning System
InSAR	Interferometric Synthetic Aperture Radar
LLNL	Lawrence Livermore National Laboratory
MWL	meteoric water line
NMO	normal move out
NRP	National Research Program
NWIS	National Water Information System
NWQL	National Water-Quality Laboratory
pmc	percent modern carbon
PSDM	pre-stack depth migration
ROE	residue on evaporation
SAR	Synthetic Aperture Radar
SLAP	Standard Light Antarctic Precipitation
TU	tritium units
USGS	U.S. Geological Survey
VPDB	Vienna Pee Dee Belemnite
VSMOW	Vienna Standard Mean Ocean Water

Well-Numbering Diagram

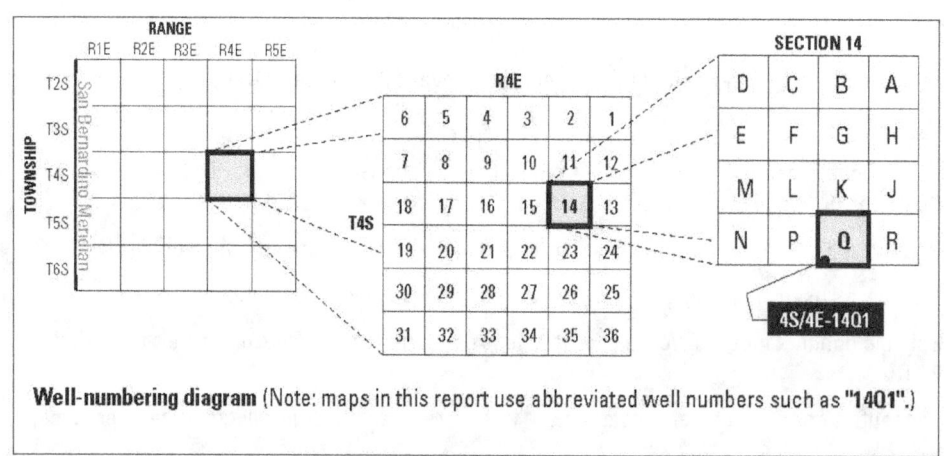

Well-numbering diagram (Note: maps in this report use abbreviated well numbers such as "14Q1".)

The Source, Discharge, and Chemical Characteristics of Water from Agua Caliente Spring, Palm Springs, California

Peter Martin, ed., with contributions by Justin Bandt, R.D. Catchings, A.H. Christensen, A.L. Flint, M.R., Gandhok, Gini Goldman, K.J. Halford, V.E. Langenheim, Peter Martin, M.J. Rymer, R.A. Schroeder, G.A. Smith, and Michelle Sneed

Executive Summary

Agua Caliente Spring, in downtown Palm Springs, California, has been used for recreation and medicinal therapy for hundreds of years and currently (2008) is the source of hot water for the Spa Resort owned by the Agua Caliente Band of the Cahuilla Indians. The Agua Caliente Spring is located about 1,500 feet east of the eastern front of the San Jacinto Mountains on the southeast-sloping alluvial plain of the Coachella Valley. The objectives of this study were to (1) define the geologic structure associated with the Agua Caliente Spring; (2) define the source(s), and possibly the age(s), of water discharged by the spring; (3) ascertain the seasonal and longer-term variability of the natural discharge, water temperature, and chemical characteristics of the spring water; (4) evaluate whether water-level declines in the regional aquifer will influence the temperature of the spring discharge; and, (5) estimate the quantity of spring water that leaks out of the water-collector tank at the spring orifice.

A gravity survey was conducted to define the thickness of the valley-fill deposits or depth to the basement complex beneath the Agua Caliente Spring area and to delineate geologic structures associated with the spring. The gravity data indicated that the Agua Caliente Spring is located within the inferred trace of the Palm Canyon fault, where the density boundaries suggest that the fault steps laterally to the west. The thickness of the valley-fill deposits is irregular along the western margin of the Coachella Valley, with a shallow buried ridge that strikes east-northeast as much as 10,000 feet away from the mountain front that appears to be a subsurface continuation of the steep ridge to the north of Tahquitz Canyon. The Agua Caliente Spring is located on the southeast flank of this buried basement ridge, where the valley-fill deposits are estimated to be 830 feet thick.

Shallow-depth seismic refraction and reflection surveys were conducted along three lines near the Agua Caliente Spring to help delineate and image geologic structures associated with the spring. Consistent with observations from nearby wells, analysis of the seismic velocity images suggests that a perched groundwater table occurs in the upper 30 feet of sediments near the spring. The seismic reflection data indicate that the basement complex is about 830 feet below land surface directly beneath the Agua Caliente Spring and that the basement complex rises from south to north, indicating the presence of a buried basement ridge to the north of the Agua Caliente Spring; this interpretation is consistent with the gravity data. The migrated seismic reflection images indicate the presence of a density contrast above the seismic interpreted depth to basement complex, which is interpreted as the contact between overlying unconsolidated valley-fill deposits and underlying indurated valley-fill deposits. The seismic interpreted contact between the unconsolidated valley-fill deposits and the indurated valley-fill deposits is about 500 feet below land surface directly beneath Agua Caliente Spring and rises to about 200 feet below land surface less than 500 feet east and north of the spring. These seismic reflection images also show disruptions in the layering and changes in the character of reflectors in the strata beneath the Agua Caliente Spring, which probably are related to the north-south trending Palm Canyon fault. Faulting of the basement complex (along the buried ridge) and indurated valley-fill deposits could provide a pathway for deep thermal water to rise from an underlying geothermal reservoir, and is the probable source of the Agua Caliente Spring.

Interferometric Synthetic Aperture Radar was used in this study to help identify ground-surface deformation and locate structures such as faults that may affect groundwater movement. Analysis of 18 interferograms representing time periods ranging from 35 to 595 days between October 2003 and September 2005 indicates that little deformation (less than 0.6 inches) occurred in the study area for the time periods represented by the interferograms. With so little deformation, none of the interferograms had sufficient contrast to provide information on the location of possible buried faults near the Agua Caliente Spring.

Historical records indicate that the Agua Caliente Spring discharge has varied from 5 to 60 gallons per minute over the past century. For this study, discharge at Agua Caliente Spring was measured by using two methods to obtain a reliable continuous record of discharge during the 2-year study period. Data collected for this study indicate that the discharge varied from a high of about 24 gallons per minute in the summer of 2005, following 2 years that had above-normal precipitation, to a low of about 9 gallons per minute in the summer of 2006, a year with below-normal precipitation. These observations suggest that the discharge of Aqua Caliente Spring is influenced by recent precipitation, although discharge data need to be collected over a period spanning multiple wet and dry cycles to establish the relation with a high degree of confidence.

Available records indicate that the temperature of the Agua Caliente Spring has been relatively constant over the past century, ranging from a low of 37.8 degrees Celsius in 1917 to a high of 42.2 degrees Celsius in 1953. Measured water temperatures at Agua Caliente Spring during this study were nearly constant, ranging from 40.7 to 41.8 degrees Celsius between April 2005 and September 2006. The temperature of the spring does not appear to be influenced by recent precipitation.

Seasonal water-quality data collected during this study and available historical data were used to define the source(s) and age(s) of water discharged by the Agua Caliente Spring, and to ascertain the seasonal and longer-term variability of chemical characteristics of the spring discharge. A large contrast in sodium fraction and pH values indicates little or no contribution from groundwater in the regional aquifer to the thermal Agua Caliente Spring. Chemical composition changed minimally in the Agua Caliente Spring during 2005–06, either seasonally or annually, indicating an absence of response to changing discharge or precipitation. Comparison with historical data indicates water quality at the spring has not changed appreciably in the last 100 years. Together, this indicates an absence of contribution to the spring from the regional aquifer and suggests an old age for the source water.

Comparison of chemical concentrations between the Agua Caliente, Fenced, and Chino Warm Springs indicates differences are much greater for major ions than for several trace elements; hence, a single common source for the geothermal water at the three sites is unlikely. Also, there are large differences between the trace element concentrations in the Agua Caliente Spring and the surrounding groundwater, which supports the inference based on major-ion concentrations that no mixing occurs between the thermal water and regional aquifer.

Temperature estimates for the geothermal reservoirs (geothermal source water) of Agua Caliente, Fenced, and Chino Warm Springs made by using an empirical relationship between sodium, potassium, and calcium concentrations and by using calculations based on aqueous equilibration with chalcedony (silica) range from 61 to 71 degrees Celsius and from 50 to 80 degrees Celsius, respectively. Both methods confirm a moderate temperature, far below the boiling point of water, for the geothermal source water for all three warm springs.

Use of dissolved-gas-concentration data yield calculated recharge temperatures of about 14 degrees Celsius for Agua Caliente Spring, 16 degrees Celsius for Fenced Spring, and 19 degrees Celsius for Chino Warm Spring. Partial loss of gas, either during sampling or by re-equilibration with soil gas as groundwater nears the surface, will cause temperature estimates based on gas concentrations to be high. The calculated recharge temperature for Chino Warm Spring probably is several degrees higher than the actual recharge temperature because excess-air data collected from the sample indicate that gasses were "stripped" from the sample during the sampling process.

Delta deuterium values range from about –70 per mil in Fenced Spring to almost –80 per mil in Agua Caliente and Chino Warm Springs. The lighter (more negative) deuterium ratios in Chino Warm and Agua Caliente Springs are consistent with an older and(or) higher-altitude source of recharge for these springs. The altitude of recharge was estimated by using deuterium data from the spring discharge and the isotopic composition of precipitation from a monitoring station on Mt. San Jacinto. The altitude of recharge was estimated to be about 7,740 feet for Chino Canyon Creek, 7,260 feet for Chino Cold Spring, 7,750 feet for Chino Warm Spring, and 7,290 feet for Agua Caliente Spring. The calculation yields a recharge altitude of about 6,100 feet for Fenced Spring; however, recharge probably was a few hundred feet higher because it is likely that evaporation has caused the isotope ratios to become less negative at this site.

Tritium is present at low concentrations in Chino Cold Spring and in a sample from the regional aquifer, indicating at least some contribution from water that is younger than 1950 (post-bomb). The complete absence of tritium at Agua Caliente Spring is consistent with the lack of mixing with groundwater in the regional aquifer.

Carbon-14 activities for samples from the Agua Caliente, Chino Warm and Fenced Springs range from 16 to 30 percent modern carbon. Calculated ^{14}C ages range from about 15,000 years before present at Agua Caliente Spring to 7,000 years before present at Chino Warm Spring. Carbon-13/12 ratios indicate some exchange with radiocarbon-dead carbonate in the soil, suggesting actual time since recharge is about 3,000 years less than these calculated ages.

Numerical models of fluid and temperature flow were developed for the Agua Caliente Spring to (1) test the validity of the conceptual model that the Agua Caliente Spring enters the valley-fill deposits from fractures in the underlying basement complex and rises through more than 800 feet of valley-fill deposits by way of a washed-sand conduit and surrounding low-permeability deposits (spring chimney) of its own making, (2) evaluate whether water-level declines in the regional aquifer will influence the temperature of discharging water, and (3) determine the source of thermal water in the perched aquifer. A radial-flow model was used to test the conceptual model and the effect of water-level declines. The observed spring discharge and temperature could be simulated if the vertical hydraulic conductivity of the spring orifice was about 200 feet per day and the horizontal hydraulic conductivity of the orifice (spring chimney) was about 0.00002 feet per day. The simulated vertical hydraulic conductivity is within the range of values reported for sand; however, the low value simulated for the horizontal hydraulic conductivity suggests that the spring chimney is cemented with increasing depth. Chemical data collected for this study indicate that the water at Agua Caliente Spring is at saturation with respect to both calcite and chalcedony, which provides a possible mechanism for cementation of the spring chimney. A simulated decline of about 100 feet in the regional aquifer had no effect on the simulated discharge of Agua Caliente Spring and resulted in a slight increase in the temperature of the spring discharge. Results from the radial-flow- and three-dimensional models of the Agua Caliente Spring area demonstrate that the distribution and temperature of thermal water in the perched water table can be explained by flow from a secondary shallow-subsurface spring orifice of the Agua Caliente Spring not contained by the steel collector tank, not by leakage from the collector tank.

Introduction

Naturally occurring hot groundwater with temperatures above 100 degrees Fahrenheit (°F) is fairly common in southern California (Moyle, 1974). Hot groundwater usually is associated with major faults and is often tapped by wells along the San Andreas, San Jacinto, and Elsinore fault zones. Hot groundwater also is tapped by wells in broad areas of the Imperial Valley and near the northwest end of the Salton Sea. While hot groundwater is fairly common, natural occurring hot springs are rare. The Agua Caliente Spring (fig. 1), owned by the Agua Caliente Band of the Cahuilla Indians, is one of only a few hot springs in the Palm Springs area and one of a handful in southern California (Dutcher and Bader, 1963).

Agua Caliente Spring has been used for recreation and medicinal therapy for hundreds of years and is currently (2008) the source of hot water for the Spa Resort owned by the Agua Caliente Band of the Cahuilla Indians. In 1958, a steel tank, open at the bottom, was constructed at the spring orifice

to collect and contain the spring's discharge. This gravel-lined tank enables the Spa Resort to have access to a clean (silt free) continuous source of the naturally heated spring water, and is still being used at present. Limited recorded historical information is available for Agua Caliente Spring. The discharge of the spring is reported to be about 25 gallons per minute (gal/min), and water temperature measurements made in 1875, 1953, and 1958 indicated that the spring's temperature varied from about 100 to 108 °F (Dutcher and Bader, 1963). Dutcher and Bader (1963) presented an explanation for the occurrence of the spring and chronicled the history of the spring before and after the installation of the steel collector tank in 1958. However, the geologic structures and hydrologic conditions that created the spring, and control its temperature and discharge, are poorly understood. The source, age, and sustainability of flow of the Agua Caliente Spring are largely unknown. How much the discharge, temperature, and chemical characteristics of the spring vary naturally, both seasonally and during wet and drought periods of several years, are not known. The hydraulic connection with regional cool-water aquifers and the spring's susceptibility to the effects of groundwater development also are not known. Because the temperature of the Agua Caliente Spring is several tens of degrees warmer than regional groundwater found at about 200 feet (ft) below land surface (bls) in the Palm Springs area, the spring could discharge a mixture of waters from different sources. Additionally, there is concern by the Agua Caliente Band of the Cahuilla Indians that a substantial quantity of spring flow may be leaking laterally from the water-collector tank into the subsurface areas surrounding the spring and beneath Indian Canyon Avenue, west of the orifice.

Purpose and Scope

The purpose of this study is to improve the understanding of the source, discharge, and chemical characteristics of the Agua Caliente Spring. The specific objectives of the study are to (1) define the geologic structure associated with the Agua Caliente Spring; (2) define the source(s), and possibly the age(s), of water discharged by the spring; (3) ascertain the natural seasonal and longer-term variability of the discharge, water temperature, and chemical characteristics of the spring water; (4) evaluate whether water-level declines in the regional aquifer will influence the temperature of the spring discharge; and (5) estimate the quantity of spring water that leaks out of the water-collector tank at the spring orifice. The information gained from this study will allow the Agua Caliente Band of the Cahuilla Indians to have a greater understanding of the hydrologic and chemical characteristics of the Agua Caliente Spring, enabling them to better preserve and manage this valuable resource.

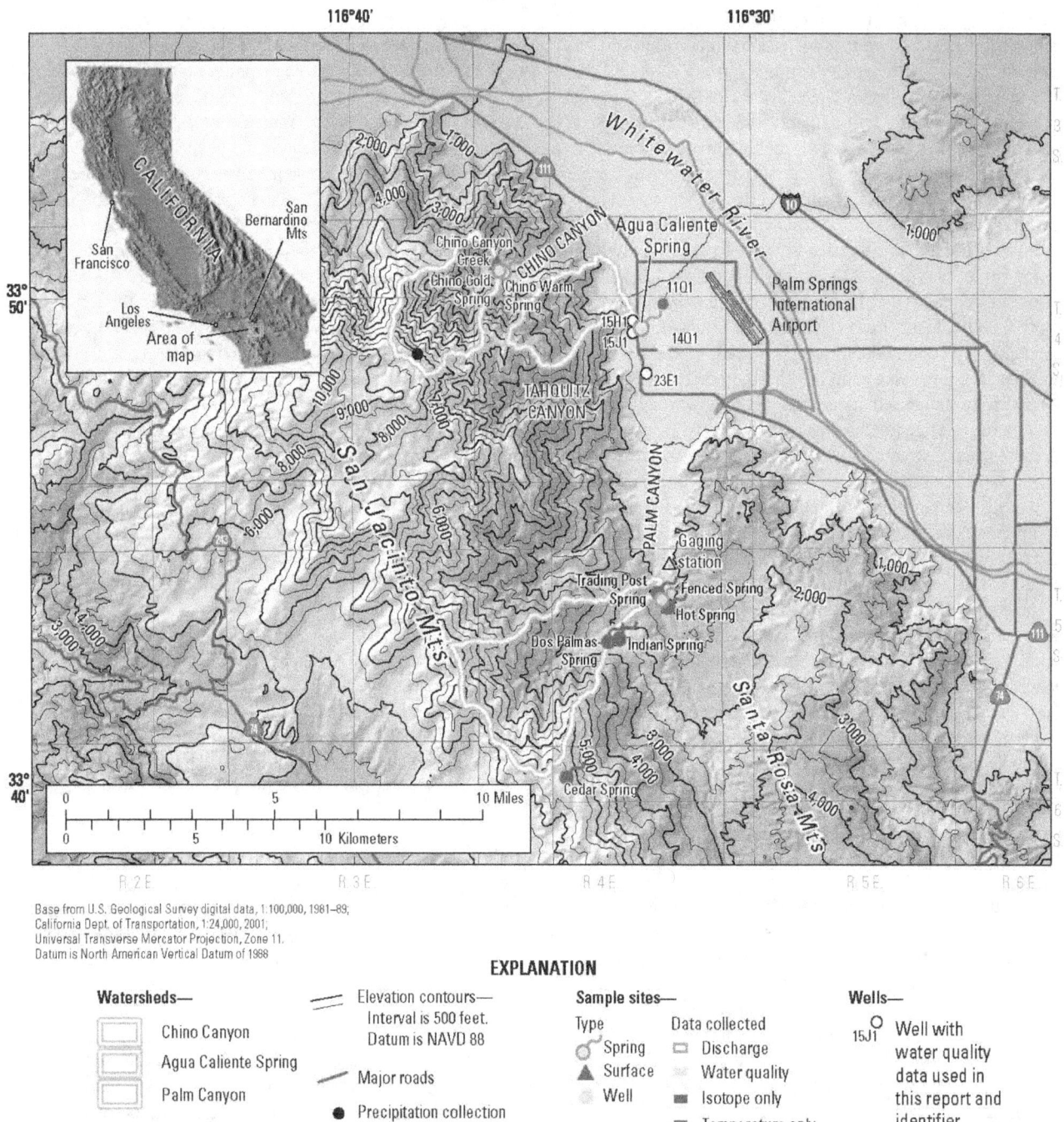

Base from U.S. Geological Survey digital data, 1:100,000, 1981–89;
California Dept. of Transportation, 1:24,000, 2001;
Universal Transverse Mercator Projection, Zone 11.
Datum is North American Vertical Datum of 1988

EXPLANATION

Watersheds—
☐ Chino Canyon
☐ Agua Caliente Spring
☐ Palm Canyon

Elevation contours—
Interval is 500 feet.
Datum is NAVD 88

Major roads

● Precipitation collection
station (Mount San Jacinto)

Sample sites—
Type   Data collected
○ Spring   ☐ Discharge
▲ Surface   ▨ Water quality
○ Well   ▬ Isotope only
              ▬ Temperature only

Wells—
15J1 ○ Well with
        water quality
        data used in
        this report and
        identifier

Figure 1. Location of Agua Caliente Spring study area, California.

In this study, geophysical data [gravity, seismic reflection and refraction, and Interferometric synthetic aperture radar (InSAR) data] were used to help define the geologic structure associated with the Agua Caliente Spring; chemical and isotopic data were used to determine the source (local versus regional groundwater-flow systems) and age (time since recharge) of water discharged by the spring; and, temperature data were simulated to estimate the amount of spring water that leaks from the water-collector tank. Seasonal differences in chemical and isotopic composition of the spring waters also were evaluated to determine if recent recharge from local groundwater flow systems is a significant component of the spring discharge and to help determine whether the spring is susceptible to drought or climatic change. This study is part of a larger study being completed by the Agua Caliente Band of Cahuilla Indians to develop a water-resource program for the spa and spring. As part of the larger study, the Agua Caliente Band of Cahuilla Indians and their contractors constructed shallow boreholes and monitoring wells near the Agua Caliente Spring. The data collected from these sites were used to help define the surficial geology and temperature distribution near the spring.

Description of the Study Area

By Gregory A. Smith and Peter Martin

Location

The Agua Caliente Spring is in downtown Palm Springs, Riverside County, California, about 120 miles (mi) east of Los Angeles (fig. 1). The Agua Caliente Spring study area includes part of the eastern slope of the San Jacinto Mountains and the northwestern part of the Coachella Valley. The Coachella Valley is a northwest-southeast trending valley that is about 65 mi long and covers about 440 square miles (mi²). The valley is drained primarily by the Whitewater River system, which discharges into the Salton Sea.

In addition to the Agua Caliente Spring, which has a temperature greater than 100°F, there are several low-temperature thermal or warm springs (less than 100°F) and non-thermal (less than or equal to the mean-annual air temperature) springs in the study area. For this study, six springs (Trading Post, Fenced, Hot, Indian, Dos Palmas, and Cedar Springs) were sampled in the Palm Canyon area to the south of Agua Caliente Spring, and two springs were sampled in Chino Canyon (Chino Cold and Chino Warm Springs) to the northwest of the Agua Caliente Spring (fig. 1). Of these

springs, Trading Post Spring, Fenced Spring, and Hot Spring in Palm Canyon, and Chino Warm Spring in Chino Canyon, are warm springs and Dos Palmas Spring and Chino Cold Spring are non-thermal springs.

Climate

The climate in the valley is typical of a desert environment located in the rain shadow of a high altitude mountain range. Average annual precipitation ranges from less than 5 inches (in.) on the valley floor to as much as 40 in. in the mountains west and north of the valley. Air temperatures range from about 120°F in the summer on the valley floor to below freezing in winter in the San Jacinto Mountains.

Precipitation and temperature for the southwestern United States were estimated for a regional analysis of groundwater recharge by Flint and Flint (2007) by using spatially distributed estimates of monthly and yearly precipitation and temperature provided by PRISM (Daly and others, 2004; http://prism.oregonstate.edu/). The PRISM data are available as monthly and yearly averages for the long-term climatic record (1895-2006) at grid spacing of about 2.5 mi. These spatially coarse grids are refined and resampled to a finer resolution of about 890 ft by using a model developed by Nalder and Wein (1998). Their model combines a spatial gradient and inverse distance squared weighting (GIDS) to monthly and yearly point data with multiple regression (Flint and Flint, 2007). A search radius of 6.2 mi was used to limit the influence of distant data. Approximately 25 PRISM grid cells were used to estimate the model parameters for precipitation and temperature for each 890 ft cell, with the closest cell having the most influence.

For this study, precipitation and temperature were extracted from the regional model developed by Flint and Flint (2007) for the watersheds upgradient of Agua Caliente, Fenced, and Chino Springs to provide a comparison for data collected for this study. It should be noted that the source area for a spring probably includes areas outside of the watershed boundary, and the watershed may not contribute any water to the spring. Geochemical and isotopic data are used later in this report to estimate the recharge temperature and average altitude of recharge for the different springs. The simulated mean annual precipitation for these watersheds was 10.5, 16.1, and 18.5 in., respectively (fig. 2). Deviation from the mean precipitation was determined for each spring to show wet and dry climatic periods. Data were collected during 2005 and 2006 for this study. Annual precipitation was above the mean during 2005, and below the mean during 2006, for all of the spring sites (fig. 2).

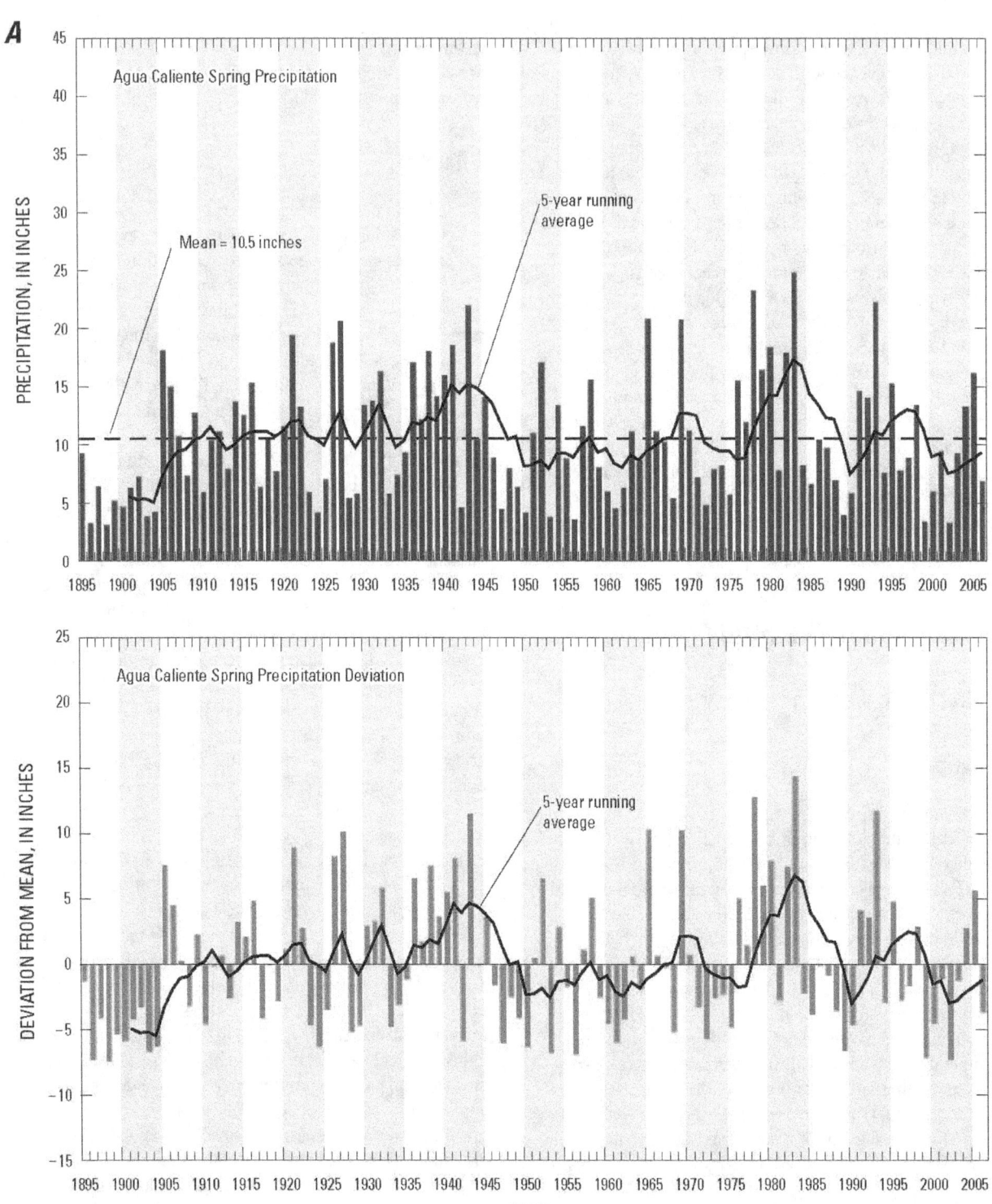

Figure 2.   Simulated precipitation and deviation from mean for the (A) Agua Caliente, (B) Fenced, and (C) Chino Springs watersheds in the Agua Caliente Spring study area, California, 1895–2006.

B

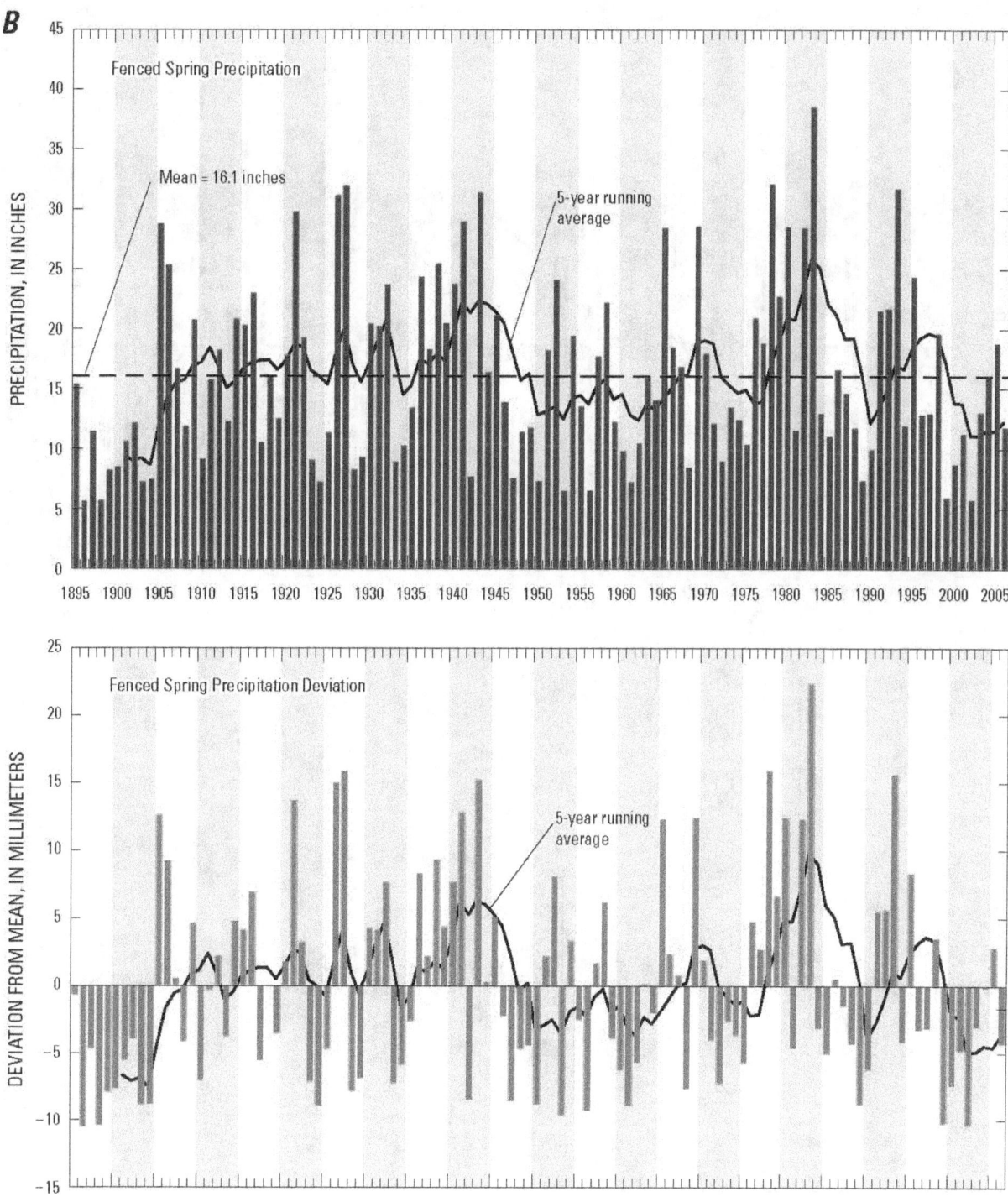

Figure 2.—Continued

C

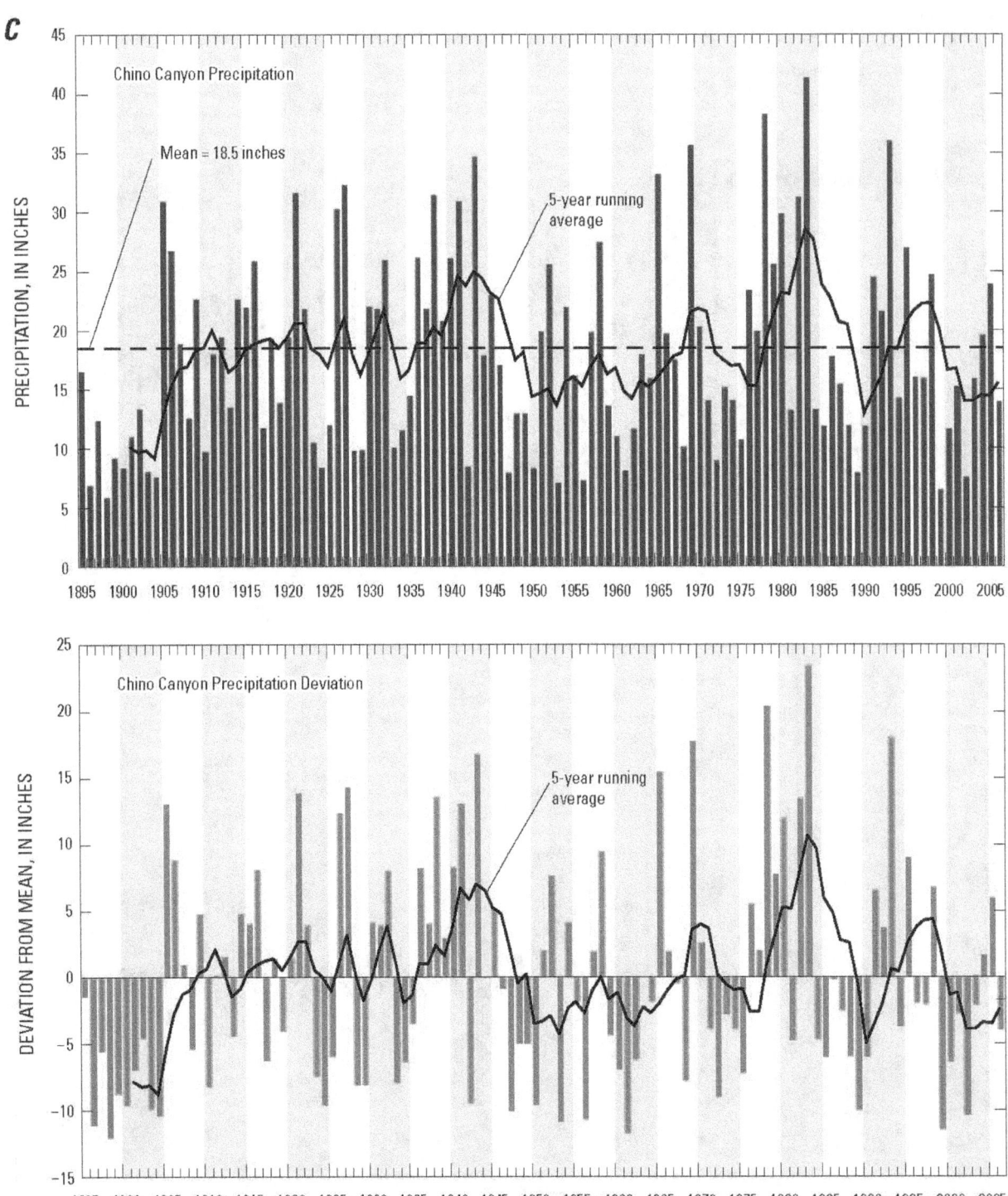

Figure 2.—Continued

History of the Agua Caliente Spring

Before the development of the spring, it is probable that a desert oasis existed at the site in which palm trees and other vegetation were sustained by the water discharged from the spring (Dutcher and Bader, 1963). There is no record of when humans first developed the spring, but the Agua Caliente Band of Cahuilla Indians have attached great significance to the spring since the earliest history of the area. Waring (1915) documented that the spring water has been used for hot "mineral baths" since the early 1900s. Brown (1923) reported that when he visited the site in 1918, the spring bubbled up in an area about 60 ft in diameter that was covered with rushes and grass and formed a pool that was partly covered by a bathhouse. Water from the spring was used by the Agua Caliente Band of Cahuilla Indians to provide hot mineral water for a bathhouse constructed at the spring and for irrigation in the vicinity of the spring (Brown, 1923). Garrett and Dutcher (1951) reported that there was a substantial bathhouse over the spring orifice and surrounding pool when they investigated the spring in 1951 (fig. 3). Overflow from the spring was discharged through buried pipes to a sump south of the spring where another pipe drained seepage from a water collector beneath the North Indian Canyon Drive (referred to as Indian Avenue by Dutcher and Bader, 1963; fig. 3). The entire flow was discharged by a pipe to a trailer park about 300 ft south of the spring where it was used in a series of "Roman baths" (Dutcher and Bader, 1963).

In 1953, the city of Palm Springs widened North Indian Canyon Drive adjacent to the spring (fig. 3). The new roadbed was several feet lower than the old road and closer to the spring, requiring the installation of a series of radiating clay drainage tiles and a central water collector to dewater the roadbed beneath the finished street (Dutcher and Bader, 1963). After the construction of the road, the spring orifice shifted to a new location, requiring the contractors to loosen the sand by jetting to restore the spring orifice to its original location (Poland and Dutcher, 1953). This event demonstrated the sensitive dynamics of the spring system.

In 1958, the Bureau of Indian Affairs started a project to stabilize the spring orifice, with technical assistance provided by the U.S. Geological Survey (USGS). There was concern that damage to the spring orifice and surrounding "chimney" might result in the permanent loss of the spring from lateral flow into the surrounding alluvium. Therefore, a steel tank, open at the bottom, was placed around the spring orifice after a 40-ft diameter hole was excavated to a depth of about 12 ft (fig. 3). The tank is about 20 ft in diameter and 10 ft deep. The tank includes a pump to take the water to baths and an overflow discharge pipe. After installation of the tank, the spring produced about 25 gal/min of silt-free water (Dutcher and Bader, 1963).

The Agua Caliente Spring is currently (2008) located under the sidewalk in front of the Agua Caliente Spa Resort on the east side of North Indian Canyon Drive. Two metal vault doors are set in the sidewalk allowing access to the spring and the collector tank that was installed in 1958. Discharge from the spring is pumped as needed to storage tanks at the Spa Resort for mineral baths, and spring discharge that is not pumped for use in the Spa Resort is pumped to a city drain along North Indian Canyon Drive.

Geohydrology of the Coachella Valley

The Coachella Valley is part of the Salton Trough, which was formed in late Cenozoic time (Loeltz and others, 1975). The Salton Trough is filled with as much as 12,000 ft of Tertiary and Quaternary sediments; the upper 2,000 ft are considered to be water bearing (California Department of Water Resources, 1979). The sediments are of continental origin, with the exception of the marine deposits of the Imperial Formation (California Department of Water Resources, 1979), and are referred to as the valley-fill deposits in this report. The valley-fill deposits tend to become more indurated (cemented) with depth and to be finer grained (contain more silt and clay) in the southern part of the Coachella Valley. In the vicinity of the Agua Caliente Spring, the sediments that compose the valley-fill deposits are, for the most part, highly permeable sands and gravels deposited principally by the Whitewater River (fig. 1), which drains the San Bernardino Mountains to the north (not shown in fig. 1). Sediments deposited by minor streams that drain the eastern slopes of the San Jacinto Mountains also are present in the valley-fill deposits, and consist of poorly sorted gravel, sand, silt, and clay.

In the Palm Springs area, the pre-Tertiary igneous and metasedimentary rocks of the San Jacinto and Santa Rosa Mountains bound the Coachella Valley to the west (fig. 4, units Mzgm and Mzg) (Jennings, 1977). These rocks are referred to as the basement complex, in this report. In general, the basement-complex rocks are of low permeability and are not considered a major water-bearing unit except where fractured or weathered.

The major faults in the Coachella Valley are part of the San Andreas fault System. These faults trend in a northwest-southeast direction and are predominantly right-lateral strike-slip faults (Jennings, 1977). The Garnet Hill fault, Banning fault, and the Mission Creek fault are part of this system, and are located along the northeastern edge of the study area (fig. 4). Dutcher and Bader (1963) postulated that the Agua Caliente Spring probably is associated with a buried, northwest-striking fault. No northwest-striking faults have been mapped in the area of the spring; however, Rogers (1966) and Jennings (1977) infer that the concealed northern extension of the north-striking Palm Canyon fault passes near the location of the spring and to the north intersects an unnamed northwest-southeast trending fault that is inferred by Jennings (1994) to be present in the western Coachella Valley adjacent to the San Jacinto and Santa Rosa Mountains (fig. 4). Gravity, seismic, and InSAR techniques were used in this study to help understand the geologic structure associated with the occurrence of the Agua Caliente Spring.

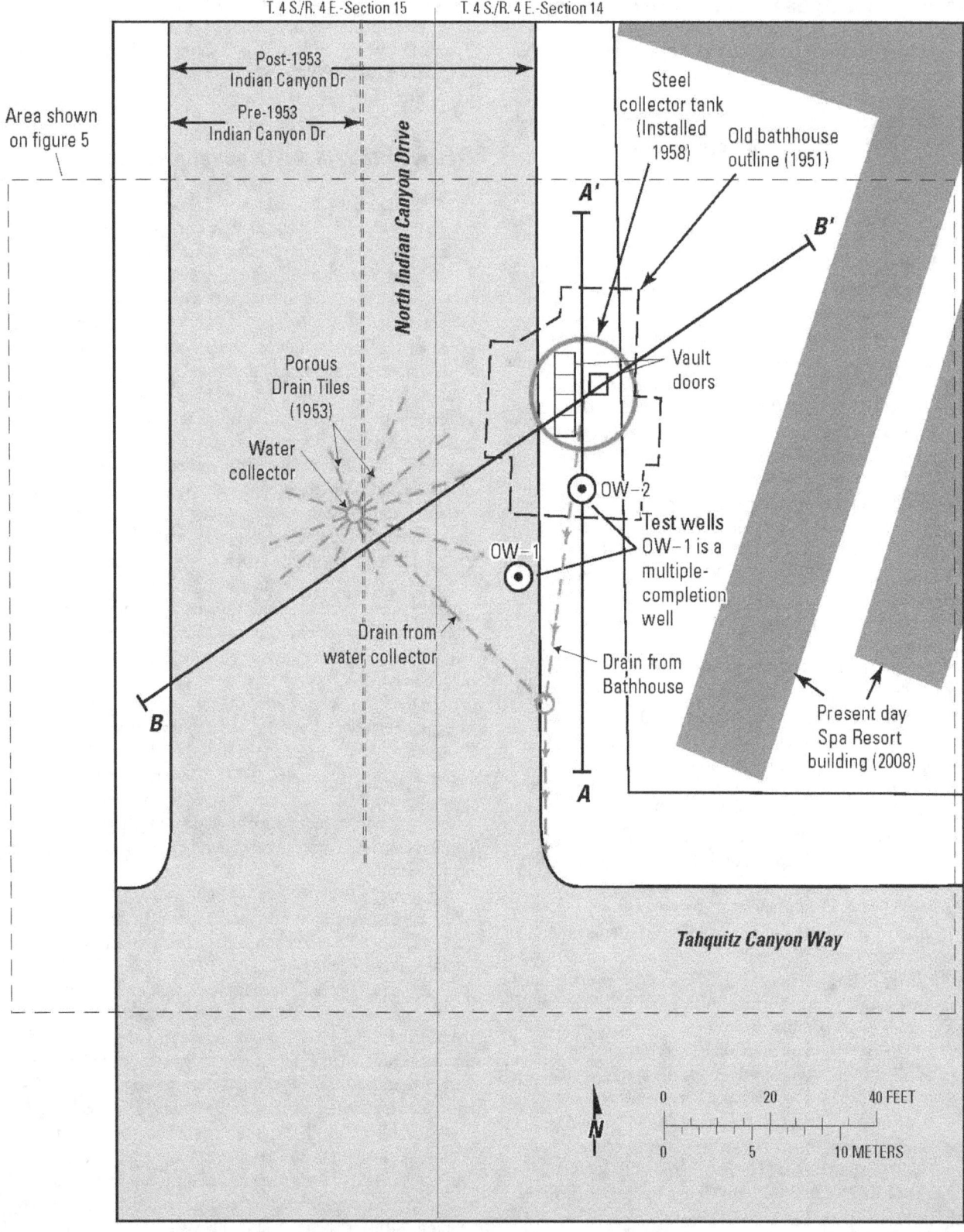

Figure 3.   The Agua Caliente Spring and associated infrastructure in the Agua Caliente Spring study area, California.

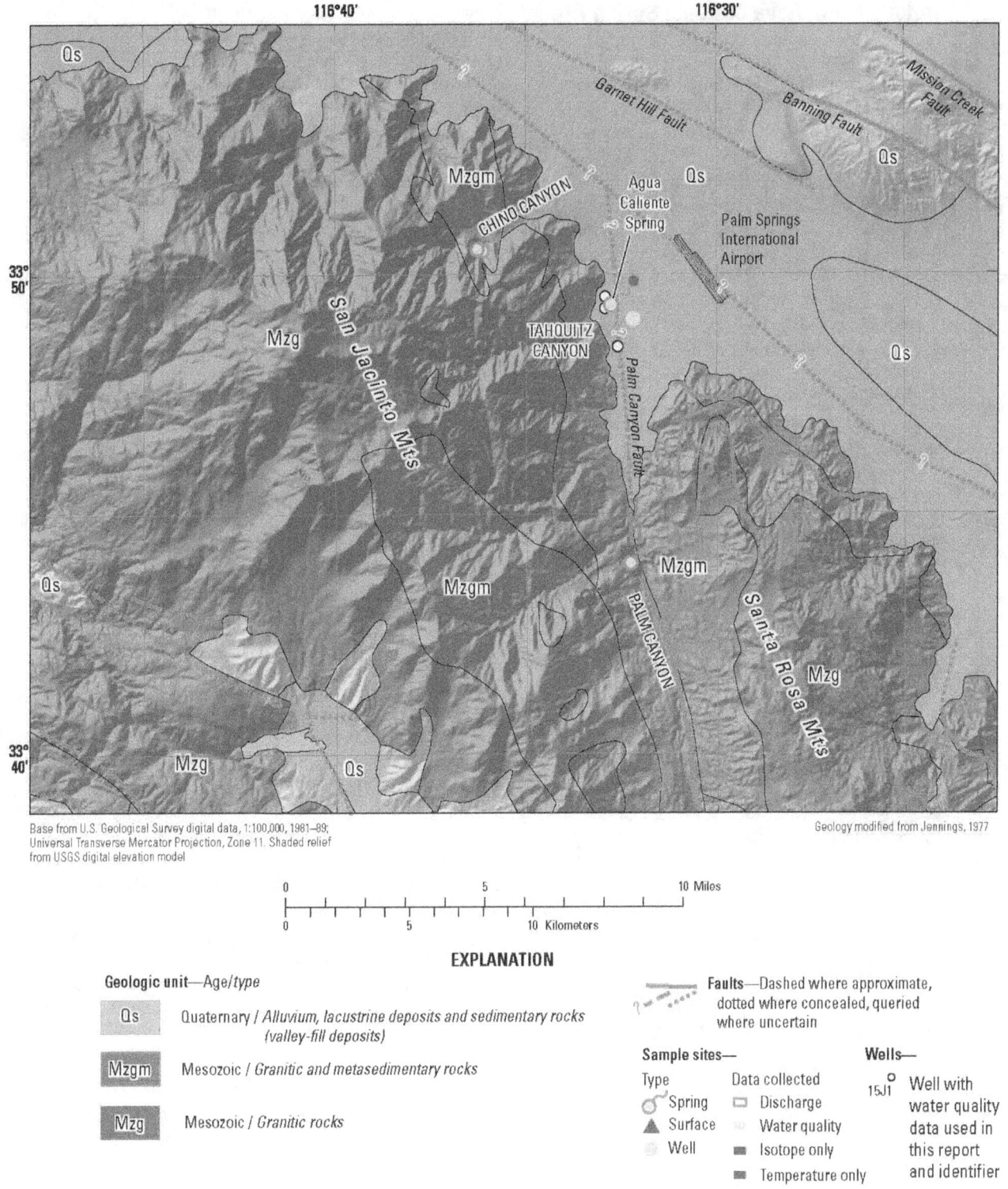

Base from U.S. Geological Survey digital data, 1:100,000, 1981–89;
Universal Transverse Mercator Projection, Zone 11. Shaded relief
from USGS digital elevation model

Geology modified from Jennings, 1977

EXPLANATION

Geologic unit—Age/type

Qs	Quaternary / *Alluvium, lacustrine deposits and sedimentary rocks (valley-fill deposits)*
Mzgm	Mesozoic / *Granitic and metasedimentary rocks*
Mzg	Mesozoic / *Granitic rocks*

Faults—Dashed where approximate, dotted where concealed, queried where uncertain

Sample sites—

Type
- Spring
- Surface
- Well

Data collected
- Discharge
- Water quality
- Isotope only
- Temperature only

Wells—

15J1 Well with water quality data used in this report and identifier

Figure 4.   Generalized geology of the Agua Caliente Spring study area, California.

Geohydrology of the Agua Caliente Spring Area

The Agua Caliente Spring is located about 1,500 ft east of the eastern front of the San Jacinto Mountains on the southeast-sloping alluvial plain of the Coachella Valley (fig. 1). The geologic log of a water well located between the spring and the mountain front (well 4S/4E-15J1, fig. 1) indicates that the alluvial-fill deposits near the spring are at least 440 ft thick. The well penetrated layers of clay, sand, and gravel, with a predominance of clay in the bottom 120 ft of the well. Well 4S/4E-15H1 (fig. 1), 750 ft northwest of the spring, was drilled to 630 ft without encountering the basement complex, indicating that valley-fill deposits may be more than 630 ft thick beneath the spring.

The regional water table measured in well 4S/4E-15H1 was about 230 ft bls in 1958. Data from other deep wells in the area confirm that the regional water table is in excess of 200 ft bls in the area of the Agua Caliente Spring. Groundwater movement in the regional aquifer is southeastward down the Coachella Valley east of Palm Springs and northeastward down Palm Canyon, south of Palm Springs (Reichard and Meadows, 1992). Beneath the Agua Caliente Spring, the direction of regional groundwater movement is eastward, toward the central part of the Coachella Valley.

Data collected from shallow wells (less than 50 ft deep) indicate that a shallow perched aquifer is present in the area surrounding the Agua Caliente Spring (Dutcher and Bader, 1963). Inspection of available geologic logs indicates that there is a low-permeability layer of silt and clay at about 30 ft; this layer impedes the vertical flow of water, creating a perched water table. Data collected in 1951 and 1959 from shallow piezometers surrounding the spring indicate that the perched water table extends at least 80 ft to the southwest (figs. 5 and 6). The presence of a water-level mound about 20 ft southwest of the orifice of the Agua Caliente Spring indicates that there is at least one additional active spring orifice in this location (figs. 5 and 6). The second spring orifice is believed to be a branch of the Agua Caliente Spring, which is not contained within the steel collector tank (fig. 6). The water-level altitudes measured in test wells OW-1 #2, OW-1 #3, and OW-2 in 2006 and 2007 were higher than the spring altitude, indicating the presence of spring discharge that is not contained within the steel collector tank (fig. 6C, table 1).

In addition, test well OW-1 #2, perforated from 25–27 ft bls, had a higher water-level altitude than test well OW-1 #3, perforated from 15-17 ft bls, indicating that the spring discharge is at or below the perforated interval of OW-1 #2 (table 1).

Installation of the steel collector tank in 1958, to stabilize the Agua Caliente Spring, involved excavating the deposits surrounding the spring to a depth of about 12 ft (Dutcher and Bader, 1963). During the excavation, two types of materials were identified in the spring mound. The bulk of the material consisted of dark green-gray, very fine sand, silt, and clay; these are referred to as peripheral deposits because they surround the spring orifice. The remainder of the material consisted of fine sand and silt that occupies nearly vertical channels, referred to as orifice deposits because they form the active and formerly active orifices of the spring deposits. Dutcher and Bader (1963) originally surmised that some type of calcareous or other cemented chimney surrounded the spring orifice to allow it to flow through more than 600 ft of unconsolidated, saturated and unsaturated valley-fill deposits; however, no evidence of a cemented chimney was found during excavation to a depth of 12 ft.

Dutcher and Bader (1963) postulated that the peripheral deposits effectively formed a low permeability "chimney" surrounding the permeable orifice deposits, greatly limiting the lateral movement of the spring water. Although these chimney deposits showed no evidence of cementation in the upper 12 ft, cementation may occur at greater depth (Dutcher and Bader, 1963). Dutcher and Bader (1963) postulated that the spring enters the alluvial sediments from fractures in the underlying basement complex. Because the spring orifice in the upper 12 ft did not show any evidence of cementation, they believed that the spring water rises through a washed-sand conduit of its own making and that the lateral movement of the spring is impeded by the low permeability, sandy-silty clay that was deposited or reworked from the alluvial deposits. Through geologic time, alluvial deposits periodically could have covered the spring orifice during flooding events. Subsequently, the spring water under pressure could have washed the fine-grained alluvial deposits away from the orifice, forming the low-permeability peripheral deposits. The present-day head in the spring system is about 200 ft higher than the regional water table.

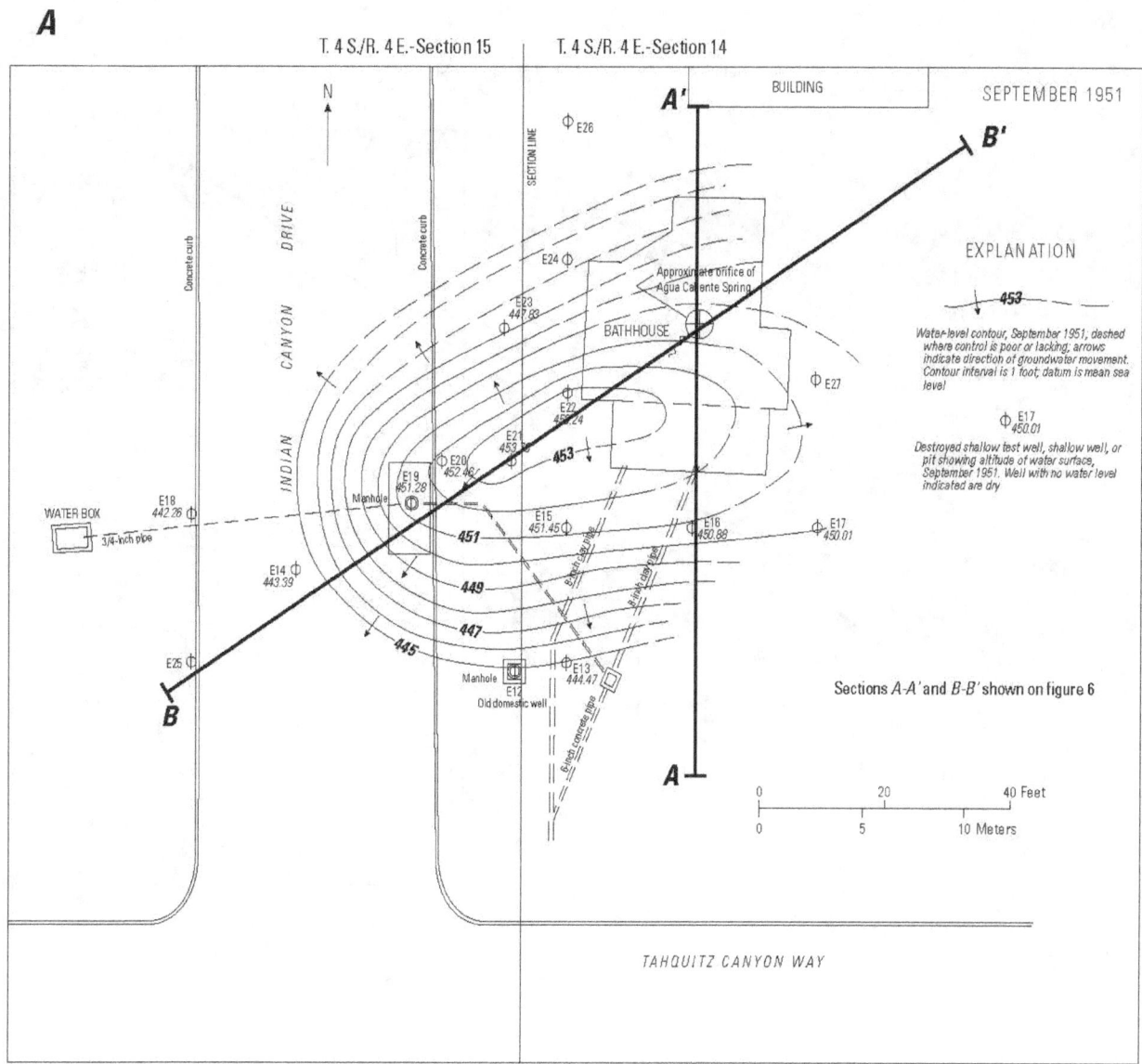

Figure 5. Water-level contours of the perched water table near the Agua Caliente Spring, California, (*A*) 1951 and (*B*) 1959.

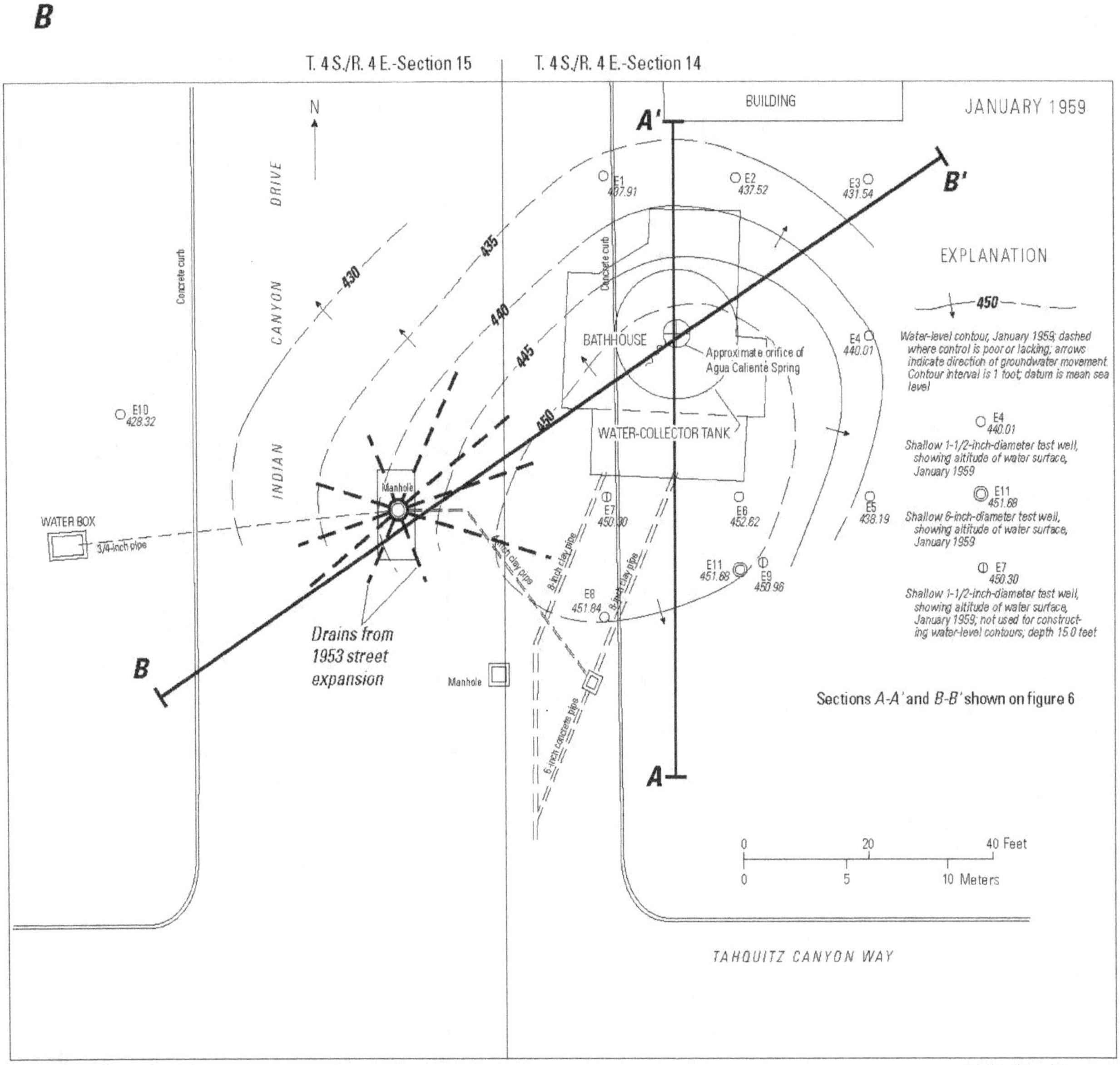

Modified from Plate 2 of Dutcher and Bader, 1963

Figure 5.—Continued

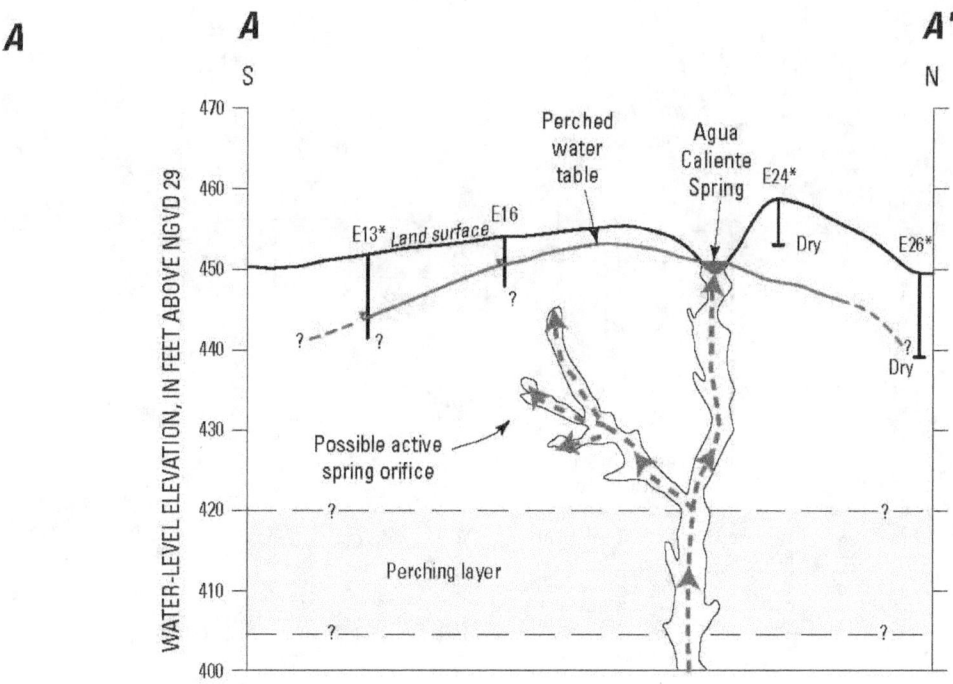

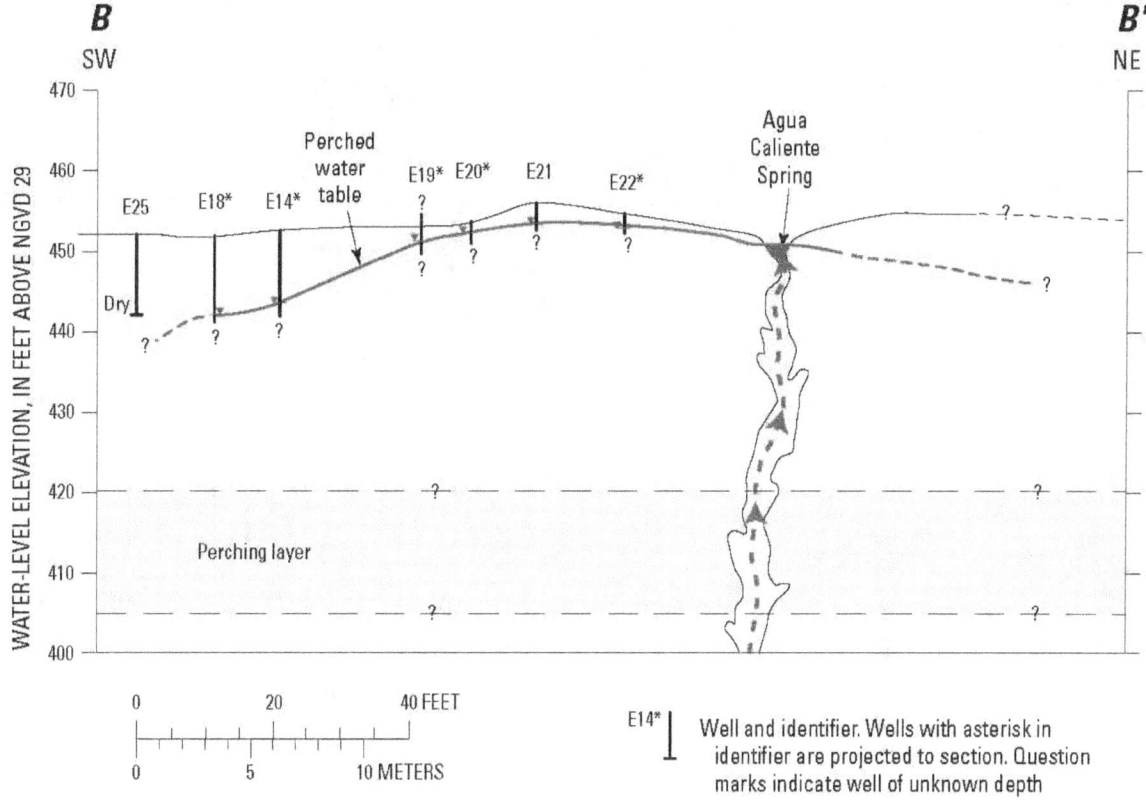

Figure 6. Geohydrologic cross section and the perched water table near the Agua Caliente Spring, (*A*) 1951, (*B*) 1959, and (*C*) 2007 of the Agua Caliente Spring study area, California.

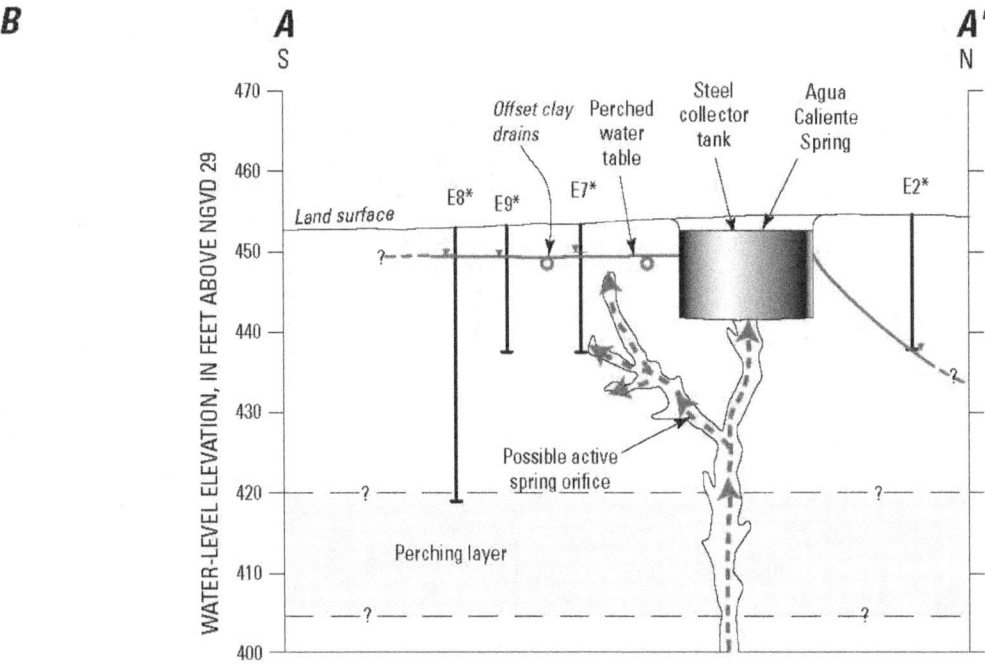

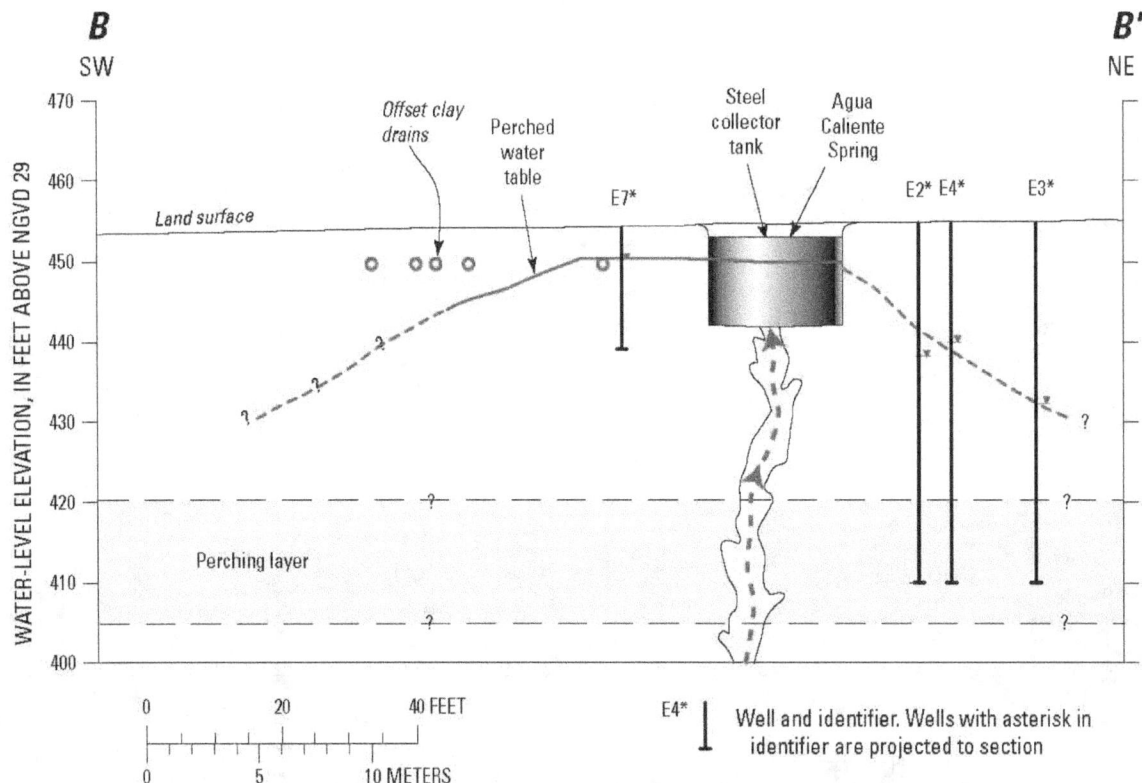

Figure 6.—Continued

C

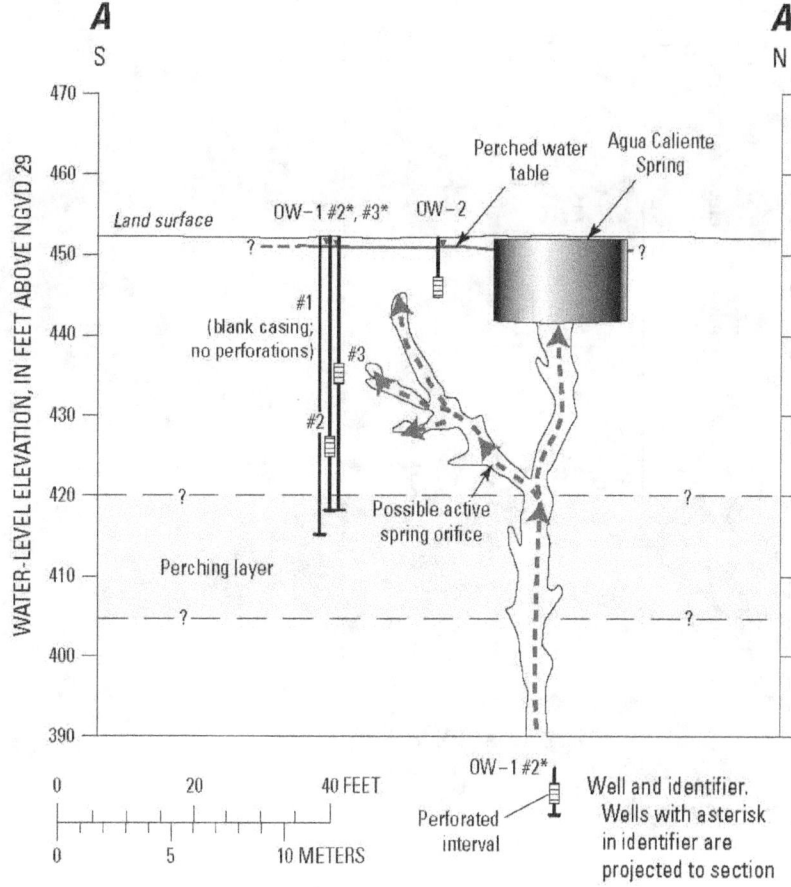

Figure 6.—Continued

Table 1. Water-level measurements in the Agua Caliente Spring, California, collector tank and wells OW–1 and OW–2, 2006–07.

[See figure 2 for well locations; land elevation, altitude of land surface, in feet above sea level, which refers to the National Geodetic Vertical Datum of 1929 (NGVD of 1929); water elevation, altitude of potentiometric surface, in feet above sea level, which refers to the National Geodetic Vertical Datum of 1929 (NGVD of 1929). **Abbreviations:** USGS ID, U.S. Geological Survey identification number; the unique number for each site in the USGS NWIS (National Water Information System) database; mm/dd/yyyy, month/day/year; ft, feet; ft BLS, depth in feet below land surface datum; F, well flowing; na, not applicable; –, no data; nc, not collected]

State well no.	USGS ID	Local identifier	Land elevation (ft)	Date (mm/dd/yyyy)	Well depth (ft)	Well perforation interval (ft)	Water depth (ft BLS)	Water elevation (ft)	Water elevation in collector tank (ft)
004S004E014ES01	334924116324301	Agua Caliente Spring	453.05	01/04/2007	na	na	2.67	na	450.38
004S004E014E001	334924116324401	OW–1 #1	452.40	–	37	na	na	na	na
004S004E014E002	334924116324402	OW–1 #2	452.40	02/10/2006	27	25–27	1.26	451.14	na
				01/04/2007	27	25–27	0.47	451.93	na
				08/01/2007	27	25–27	F	nc	na
004S004E014E003	334924116324403	OW–1 #3	452.40	02/10/2006	17	15–17	1.72	450.68	na
				01/04/2007	17	15–17	0.96	451.44	na
				08/01/2007	17	15–17	0.68	451.72	na
004S004E014E004	334923116324401	OW–2	452.64	02/10/2006	7	5–7	2.56	450.08	na
				01/04/2007	7	5–7	1.61	451.03	na
				08/01/2007	7	5–7	1.77	450.87	na

Investigating the Geologic Structure of the Agua Caliente Spring

The presence of the Agua Caliente Spring about 1,500 ft east of the San Jacinto Mountains is surprising because data from water wells indicate that the valley-fill deposits are at least 600 ft thick and the regional water table is in excess of 200 ft bls in the vicinity of the spring. Prior to this study, limited deep subsurface geologic information was available to evaluate the potential for buried bedrock structures or other explanatory factors for the occurrence of the spring. Gravity, seismic, and InSAR techniques were used in this study to help understand the geologic structure associated with the occurrence of the Agua Caliente Spring.

Gravity Survey of the Agua Caliente Spring Area

By Victoria E. Langenheim and Allen H. Christensen

A gravity survey was conducted to define the thickness of the valley-fill deposits, or depth to the basement complex, beneath the Agua Caliente Spring area and to delineate geologic structures associated with the spring (fig. 7). The study area for the gravity survey extends from the gaging station in Palm Canyon north to Desert Angel, and from the Palm Springs International Airport west to the base of the aerial tramway in Chino Canyon (an area approximately 6 mi by 6 mi). This gravity survey did not include defining the thickness of the valley-fill deposits in Chino and Tahquitz canyons, but focused on the thickness of the valley-fill deposits in Palm Canyon and beneath Palm Springs.

Gravity, geologic maps, and water wells were the primary data sets used to define the thickness of the valley-fill deposits. Aeromagnetic data were examined and deemed not useful because the basement-complex rocks and the valley-fill deposits are not sufficiently magnetic to produce measurable anomalies. The valley-fill deposits (fig. 4, unit Qs) are made up of locally derived Quaternary sediments and rest on crystalline rocks of the basement complex that are Mesozoic and older (fig. 4, units Mzgm and Mzg). The large density contrast between the valley-fill deposits and the crystalline basement complex (here averaging 550 kilograms per cubic meter [kg/m^3]) makes determining the thickness of the valley-fill deposits a good candidate for study by gravity methods.

Data Sets

Gravity Survey and Reduction

Data were collected at 252 new gravity stations along east-west traverses across Palm Canyon and the area of the Agua Caliente Spring. These data supplemented data previously collected at about 70 stations (Biehler and others, 2004; V.E. Langenheim, USGS, unpub. data). Gravity data were reduced using the Geodetic Reference System of 1967 (International Union of Geodesy and Geophysics, 1971) and referenced to the International Gravity Standardization Net 1971 gravity datum (Morelli, 1974, p. 18). Gravity data were reduced to isostatic anomalies using a reduction density of 2,670 kg/m^3 and include earth-tide, instrument drift, free-air, Bouguer, latitude, curvature, and terrain corrections (Telford and others, 1976). An isostatic correction using a sea-level crustal thickness of 16 mi and a mantle-crust density contrast of 400 kg/m^3 was applied to the gravity data to remove the long-wavelength gravitational effect of isostatic compensation of the crust due to topographic loading. The data were gridded at a spacing of 820 ft, roughly the spacing of gravity stations along the detailed profiles, by using a minimum curvature algorithm. The resulting gravity field is termed the isostatic residual gravity anomaly.

Terrain corrections were computed to a radial distance of 104 mi and involved a three-part process: (1) Hayford-Bowie zones A and B with an outer radius of 223 ft were estimated in the field with the aid of tables and charts; (2) Hayford-Bowie zones C and D with an outer radius of 1,936 ft were computed using a 100-ft digital elevation model; and (3) terrain corrections from a distance of 1,936 ft to 104 mi were calculated by using a digital elevation model and a procedure by Plouff (1977). Total terrain corrections for the stations measured for this study ranged from 0 to 18.4 milli-Galileo (mGal), averaging 9.1 mGal. If the error resulting from the terrain correction is considered to be 5 to 10 percent of the total terrain correction, the largest error from the terrain correction expected for the data is 1.8 mGal. However, the error resulting from the terrain correction is small (less than 0.5 mGal) for most of the stations because there are minimal changes in relief between most of the stations.

Geologic Maps

Geologic data from three maps were used in the gravity study. The geologic map of the Palm Springs 7.5-minute quadrangle (Dibblee, 2004) was used for most of the study area, with additional information from the Palm Springs 15-minute quadrangle (Dibblee, 1981a) and adjacent Idyllwild quadrangle (Dibblee, 1981b). These geologic data were used primarily to delineate the boundary between the valley-fill deposits and the basement complex.

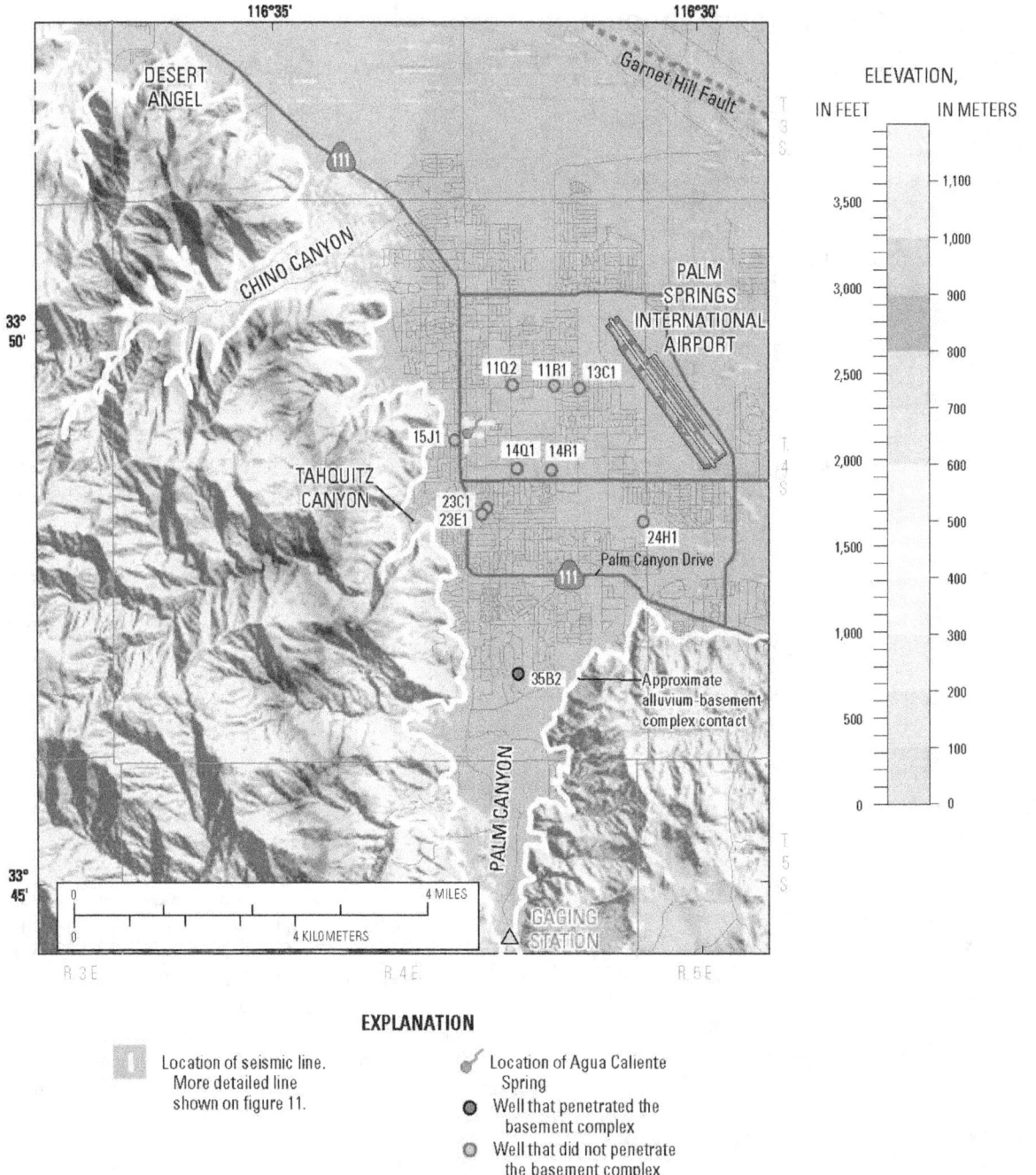

Figure 7. Location of gravity survey and wells with geologic data in the Agua Caliente Spring study area, California.

EXPLANATION

Location of seismic line. More detailed line shown on figure 11.

Location of Agua Caliente Spring

Well that penetrated the basement complex

Well that did not penetrate the basement complex

Wells

A set of ten water wells (table 2) with driller's logs and well depths was available to constrain the gravity interpretations. Only one of these wells (4S/4E-35B2) penetrated the entire thickness of the valley-fill deposits (735 ft).

Gravity Field

The gravity field of the study area (here expressed as the isostatic residual gravity field) is complex, and mostly reflects the large density contrast between the dense basement complex and the lower density valley-fill deposits (fig. 8). The most prominent features on the gravity map are the high gravity values (greater than –12 mGal) that coincide with basement exposures in the San Jacinto Mountains and the very low gravity values (less than –28 mGal) coincide with the thick valley-fill deposits of the Coachella Valley. An oil test well located about 3 mi east of the northeast corner of our study area reached 7,474 ft bls without penetrating the basement complex (Proctor, 1968), illustrating the great depth to the basement complex beneath Coachella Valley.

Superposed on the pronounced northwest-striking gravity gradient between the highs over the crystalline basement and the lows of the deep Coachella Valley basin is a 4 to 6 mGal low in Palm Canyon, corresponding to gravity values of –14 to –20 mGal (fig. 8). The axis of this low strikes north to the latitude of Palm Canyon Drive where it curves to a more northeasterly strike. The Agua Caliente Spring is located on the southeast flank of a local gravity high.

The gravity field was analyzed to define the structural setting of the Agua Caliente Spring. The automated method of Blakely and Simpson (1986) was used to define where changes in rock density are located over a short distance, such as density contrasts caused by faults. Places where the gravity field changes the most (maximum horizontal gradient, white dots on figure 8) marks steeply dipping contacts between rocks of differing densities (density boundaries), such as steep contacts between the dense basement complex and lower density valley-fill deposits. This method uses the gridded data set and will also locate horizontal gravity-gradient maxima in low-gradient areas that may not be well constrained by gravity measurements or reflect faulting. Density boundaries form a northwest-trending alignment, parallel to the strike of the Garnet Hill and Banning faults of the San Andreas system that passes near the Palm Springs International Airport. The northwest-trending density boundaries do not pass through the Agua Caliente Spring. Instead, density boundaries highlight a northward continuation of the Palm Canyon fault, along the eastern edge of the San Jacinto Mountains, that passes through the Agua Caliente Spring. A density boundary also follows the eastern margin of Palm Canyon, projecting about 10,000 to 13,000 ft northeastward beyond the northernmost crystalline outcrops toward the airport. No known northeast-striking faults have been mapped in this area, so this boundary is interpreted as representing a steep, buried canyon wall.

Table 2.   Location, well depth, and depth to basement complex for ten wells in the Agua Caliente Spring area, California.

[See figure 7 for site locations, State well number, see well-numbering diagram in text. **Abbreviations:** °, degree; ft, feet; ft BLS, depth in feet below surface datum; –, no data]

State well no.	Latitude (°North)	Longitude (°West)	Well depth (ft BLS)	Measured depth to basement complex (ft)	Calculated depth to basement complex (ft)
004S004E11Q002	33.8306	116.5358	948	–	1,993
004S004E11R001	33.8304	116.5275	737	–	1,466
004S004E13C001	33.8301	116.5228	912	–	1,376
004S004E14R001	33.8176	116.5286	813	–	1,521
004S004E14Q001	33.8179	116.5354	980	–	1,262
004S004E15J001	33.8224	116.5466	438	–	637
004S004E23C001	33.8124	116.5405	425	–	1,146
004S004E23E001	33.8118	116.5411	488	–	962
004S004E24H001	33.8092	116.5104	1,025	–	1,062
004S004E35B002	33.7859	116.5354	730	735	750

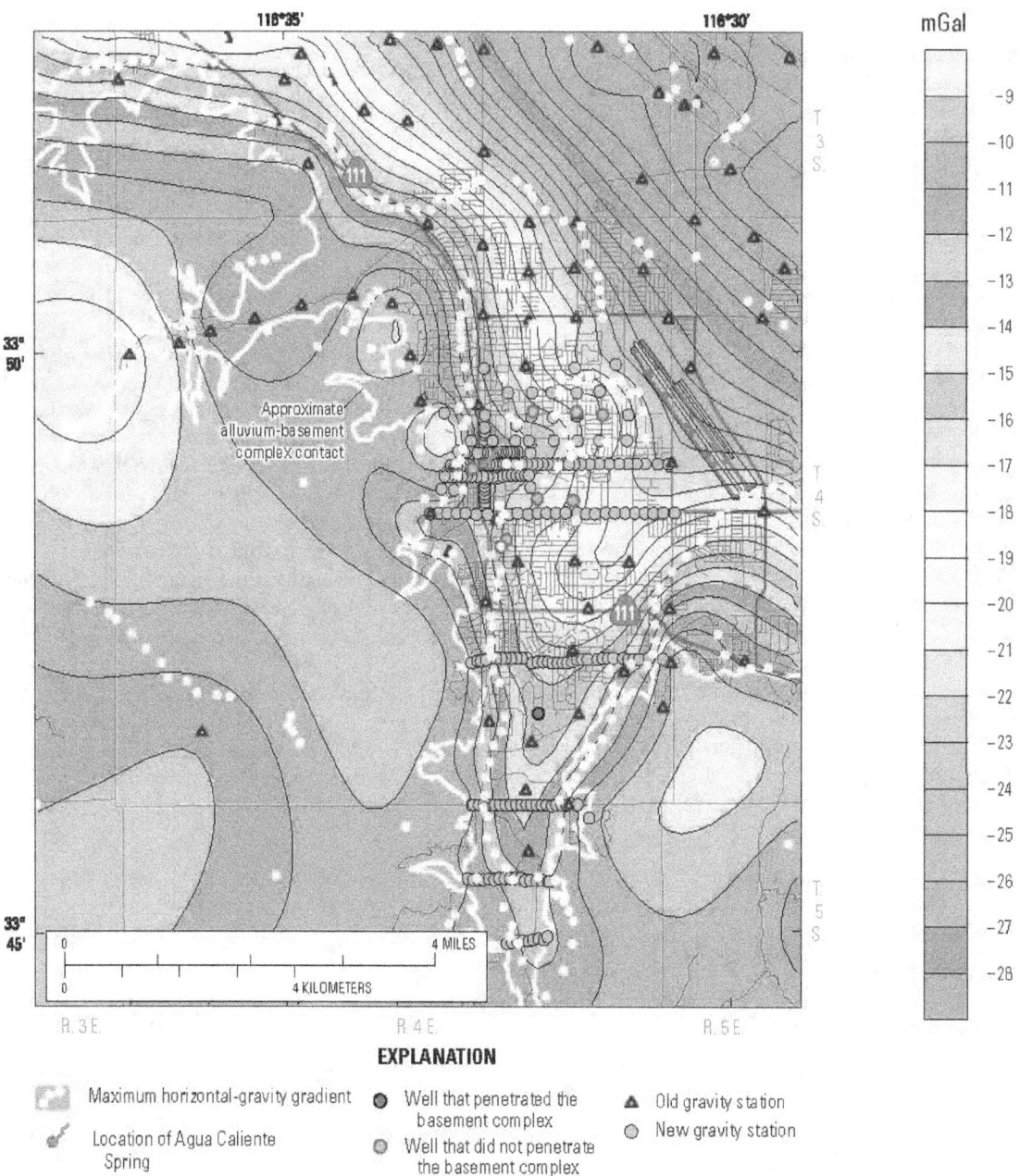

Figure 8.   Isostatic residual gravity field and location of density boundaries in the Agua Caliente Spring study area, California.

Computation Method for Modeling the Thickness of the Valley-Fill Deposits

The thickness of the valley-fill deposits (or depth to the basement complex) throughout the study area was estimated using the method of Jachens and Moring (1990), modified slightly to permit inclusion of constraints at points where the thickness of the valley-fill deposits is known from direct observations in boreholes. An initial estimate of the 'valley-fill deposits gravity anomaly' is made by passing a smooth surface through the gravity values at stations where the basement complex rocks crop out (initial estimate of the 'basement gravity field') and subtracting this from the isostatic residual gravity field. This represents only the initial estimate because the gravity values at points on basement complex that lie close to the valley-fill deposits are influenced by the lower density valley-fill deposits and are, therefore, lower than they would be if the valley-fill deposits were not present. To compensate for this effect, the initial 'valley-fill deposits gravity anomaly' is used to calculate an initial estimate of the thickness of the valley-fill deposits, and the gravity effect of these valley-fill deposits is calculated at all of the basement gravity stations. A second estimate of the 'basement gravity field' is then made by passing a smooth surface through the basement gravity values corrected by the valley-fill effect and the process is repeated to produce a second estimate of the thickness of the valley-fill deposits. This process is repeated until further steps do not result in significant changes to the modeled thickness of the valley-fill deposits, usually in five or six steps.

The 'valley-fill deposits gravity anomaly' was converted to thickness of the valley-fill deposits by using an assumed density contrast that varies with depth (table 3) between the sedimentary deposits that make up the valley-fill deposits and the underlying basement complex. This density-depth relationship was slightly modified from that derived from the Banning Pass area north of the Agua Caliente Spring study area (Langenheim and others, 2005) by comparing the 'valley-fill deposits gravity anomaly' with the thickness of the valley-fill deposits identified at well 4S/4E-35B2 (table 2). The resulting density contrast of 550 kg/m³ for the valley-fill deposits in the upper 656 ft (200 meters [m]) is reasonable for Quaternary continental deposits overlying Mesozoic and older crystalline rocks. The reasonableness of this selection of density contrast was further tested by examining the 'basement gravity field' for any indications of local anomalies at the sites where wells penetrated the basement complex, and the solution was forced to honor those data.

Gravity Results

The gravity inversion resulted in a calculated thickness of the valley-fill deposits, or depth to basement complex, that ranges from 0 ft in the mountains to the west of Agua

Table 3.   Assumed density contrast with depth in the Agua Caliente Spring area, California.

[Abbreviations: ft, feet; m, meters; BLS, below land surface datum; kg/m³, kilograms per cubic meter; >, greater than]

Depth range (ft BLS)	Depth range (m BLS)	Density contrast (kg/m³)
0–656	0–200	−550
656–1,968	200–600	−360
1,968–4,920	600–1,500	−300
>4,920	>1,500	−230

Caliente Spring to about 7,000 ft in the Coachella Valley to the northeast of the spring (fig. 9). The porosity and permeability of the valley-fill deposits undoubtedly decrease between the surface and depths of thousands of feet because of increased compaction and cementation with depth. Therefore, although the depth to basement complex is great in the northeastern part of the study area, the thickness of permeable water-bearing deposits likely is significantly less than the total calculated thickness of the valley-fill deposits.

The calculated thickness of the valley-fill deposits at the one place where a well (4S/4E-35B2) penetrated the entire thickness of the valley-fill deposits agrees with the observed thickness to within 20 ft (table 2), which is expected because the solution was constrained to honor this value. The lack of perfect agreement at this site reflects the spatial averaging. Another measure of the reliability of the solution can be obtained by comparing the calculated thicknesses with the total well depths at those wells that did not penetrate the basement complex. For all nine of these wells, the calculated thickness of the valley-fill deposits is greater than the total well depth, as it should be (table 2). Basic uncertainties in the gravity data imply that the best resolution that can be expected, even in areas of good gravity coverage, is about + 50 ft, and resolution is likely less in areas of poor gravity coverage or in areas far from either the basement outcrop or control points where wells penetrated the basement complex. Also, because the calculations were performed on grid cells 820 ft on a side, the results represent averages of the thickness of the valley-fill deposits over this cell size. Variations of the thickness of the valley-fill deposits over distances less than a cell-dimension are not resolved. Finally, gravity data reflect the average shape of the causative body (in this case, the thickness of the valley-fill deposits), and the averaging becomes more pronounced farther from the source where the measurements are taken. As a result, places where the valley-fill deposits are the thickest are subject to higher degrees of averaging, and thus appear smoother than areas where the valley-fill deposits are thinner.

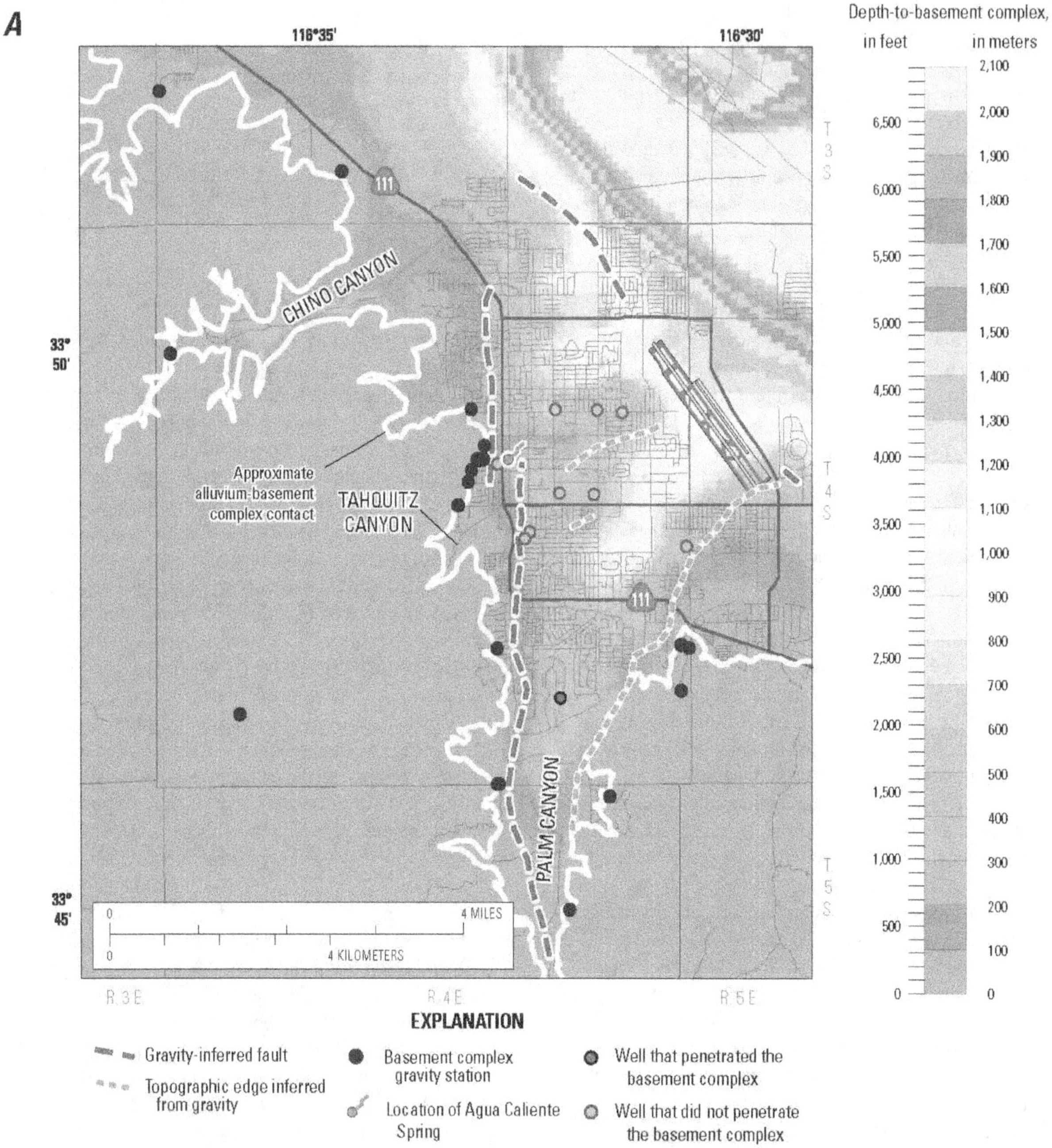

A

Figure 9.   The (*A*) thickness of the valley-fill deposits and (*B*) altitude of the top of the basement complex calculated from gravity measurements in the Agua Caliente Spring study area, California. Click on figure 9*B* to see and control animation showing the altitude of the top of the basement complex.

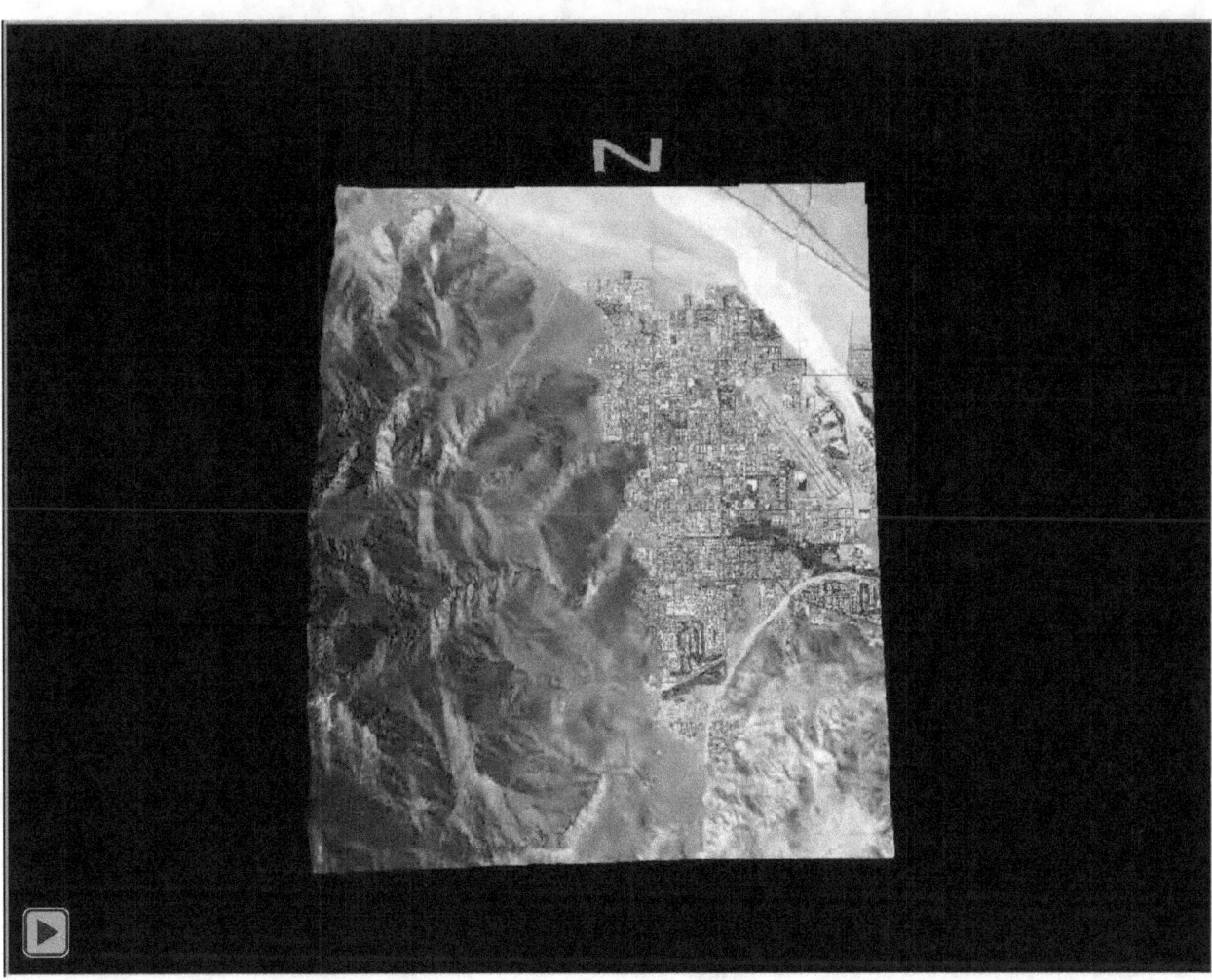

Figure 9.—Continued

Southeast of the Agua Caliente Spring, the valley-fill deposits fill a wide trough that appears to be the buried continuation of Palm Canyon (fig. 9A). The depth of the canyon increases to as much as 3,000 ft north of Palm Canyon Drive (Highway 111). The thickness of the valley-fill deposits is irregular along the western margin of the Coachella Valley, with a shallow buried ridge that strikes east-northeast as much as 10,000 ft away from the mountain front. The valley-fill deposits are estimated to be about 830 ft thick beneath Agua Caliente Spring, which is located on the southeast flank of this buried basement ridge that appears to be the subsurface continuation of the steep ridge to the north of Tahquitz Canyon (fig. 9A).

To help visualize the basin geometry in the Agua Caliente Spring area, an animation of the altitude of the top of the basement-complex was prepared (fig. 9B). The altitude of the top of the basement complex was calculated by subtracting the modeled thickness of the valley-fill deposits at each gravity grid from the average land-surface altitude at that grid. The animation allows the viewer to fly over the ridges and valleys of the shaded-relief altitude of the top of the basement complex. A light blue line indicates the location of the Agua Caliente Spring in the animation, and the dark blue marker at the top of this line represents land surface at the spring. The altitude of the top of the basement complex beneath the Agua Caliente Spring is about 450 ft above sea level (fig. 9B). As shown in the animation, there is a prominent basement ridge to the north of the Agua Caliente Spring.

The gravity data suggest that the Palm Canyon fault continues to the north of its mapped trace along the eastern margin of the San Jacinto Mountains (fig. 9A). The Agua Caliente Spring is located within the inferred trace of the Palm Canyon fault, where the density boundaries suggest that the fault steps to the west. Faulting of the basement complex along the buried ridge would allow deep thermal water to quickly rise into the overlying valley-fill deposits from an underlying reservoir of deep geothermal water, which is the probable source of the Agua Caliente Spring (fig. 10). The gravity data and the basin geometry do not support a northwest-striking fault that projects toward the Agua Caliente Spring, as proposed by Dutcher and Bader (1963). If such a fault exists, it does not have much vertical offset (less than 30 ft).

Seismic Refraction and Reflection Surveys of the Agua Caliente Spring Area

By Michael J. Rymer, Rufus D. Catchings, Mark R. Goldman, Peter Martin, and Gini Gandhok

In June 2006, the USGS Seismic Imaging Group (SIG) acquired high-resolution shallow-depth combined seismic reflection and refraction data in the vicinity of Agua Caliente Spring in downtown Palm Springs, California, to help delineate and image geologic structures associated with the spring. Seismic data were acquired along three seismic

profiles, referred to as Line 1, Line 2, and Line 3 (fig. 11). Line 1 was oriented north-south along the eastern edge of Indian Canyon Drive and extended approximately 2,000 ft, centered on the spring (fig. 11). Line 2 was about 1,800 ft long and was oriented east-west along the northern edge of Tahquitz Canyon Way and also was centered on the spring (fig. 11). Lines 1 and 2 intersect near their midpoints. Line 3 was about 1,800 ft long and was oriented east-west along the northern edge of Andreas Road (fig. 11). Line 3 intersected with Line 1 near the northern end of Line 1 and was parallel with Line 2, separated by about 650 ft.

Data Acquisition and Processing

The seismic data were acquired by using two linked multi-channel Geometrics Strataview™ RX-60 seismographs with a total of 120 channels. Seismic sources were generated by an accelerated weight drop (AWD) that propelled a steel rod onto a metal plate that was placed on the street surface. The seismographs were coupled to the ground (asphalt pavement) via Mark Products L-40A™ single-element, 40-Hertz (Hz) geophones. Both the seismic sources (shot point locations) and the geophones were spaced at 16-ft intervals over the seismic arrays, with the geophones and seismic sources laterally offset by 3.28 ft (1 m). The resulting seismic data were recorded for 2 seconds (s) at a 0.5-millisecond (ms) sampling rate without acquisition filters. To account for the effects of elevation and varying geophone and shot point locations, each geophone and shot point location were surveyed using a differential Global Positioning System (GPS). The sites were surveyed using accepted Real-Time Kinematic surveying techniques using two GPS receivers with real-time communication using radio modems for differential corrections (Morton and others, 1993). The absolute difference in elevation measured between stations is ± 0.008 ft, based on manufacture specifications (Buick, 2006).

Data were acquired in a manner that allows both refraction and reflection images to be produced. First-arrival refractions were inverted for velocity structure to develop P-wave velocity images along each seismic line using a modified version of the tomographic inversion code of Hole (1992). The data were inverted using a 16-ft by 16-ft grid along the profile to the depth of ray coverage. A first-arrival refracted arrival on each seismogram from each seismic source was inspected and measured, resulting in a total of 66,000 to 72,000 seismograms per seismic line (greater than 200,000 seismograms total). Unfortunately, because there was a high level of cultural noise, it was not possible to determine first-arrival refractions for all 200,000 seismograms, particularly for seismograms that were far from the source. The limited range for first-arrival refractions made it impossible to determine refraction velocities below about 150 ft bls. To produce stacked reflection images (see discussion below) the seismic velocities (depth-extended velocities) were estimated at depths below about 150 ft bls based on the lithologies observed in nearby boreholes and from the normal move out (NMO) of reflected arrivals.

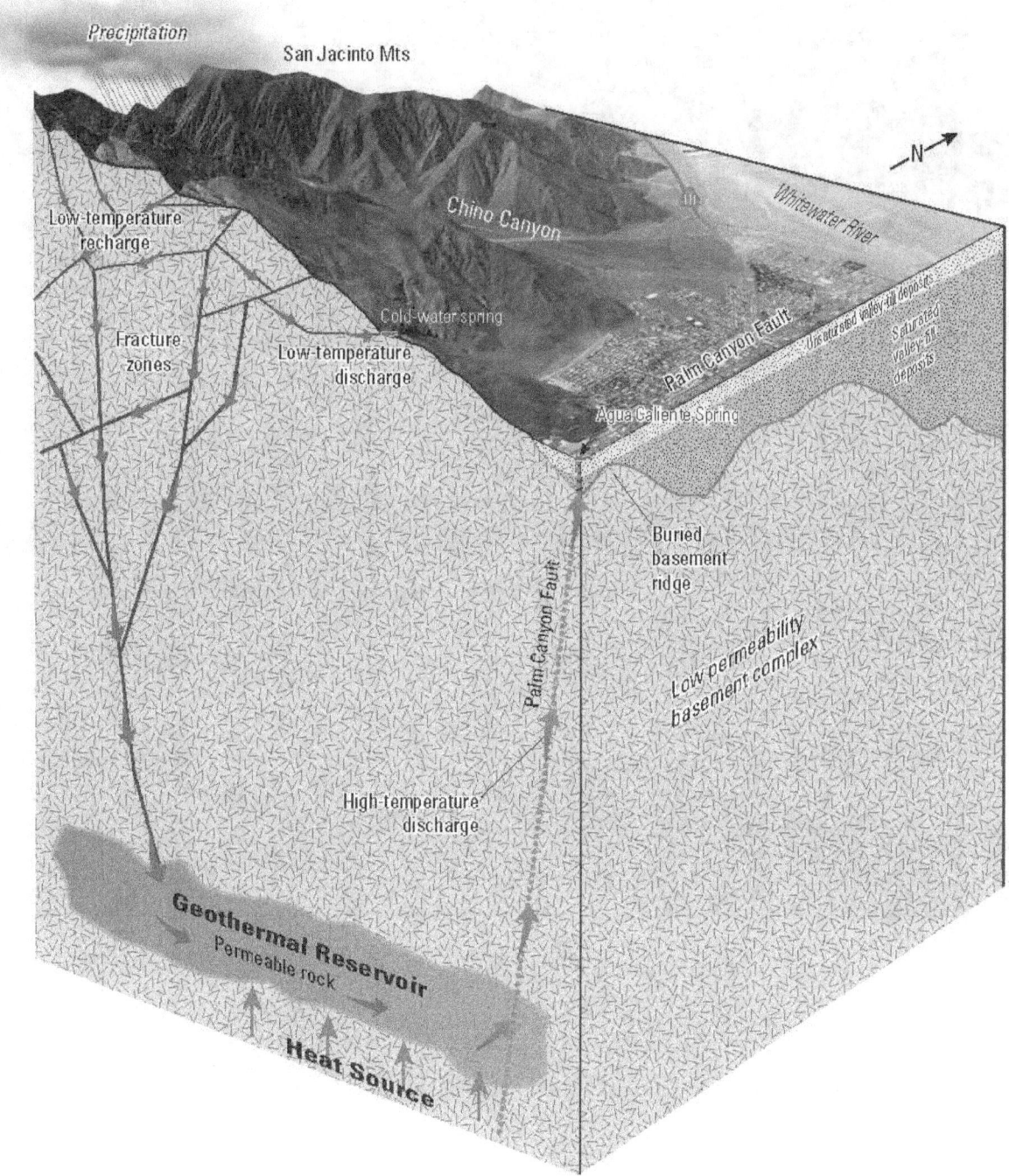

Figure 10.   The source of water to the Agua Caliente Spring in the Agua Caliente Spring study area, California.

Base from Google Maps

EXPLANATION

Location of seismic line.
General location
shown on figure 7.

0   1,000 FEET

0   300 METERS

Figure 11.   Location of seismic Lines 1, 2, and 3 in the Agua Caliente Spring study area, California.

Reflection images were developed using the seismic data and appropriate reflection processing schemes. The seismic processing procedures used for this study were similar to those outlined by Brouwer and Helbig (1998). The reflection data were processed on a Sun Sparc 20 ™ computer using the interactive seismic reflection processing software package, ProMAX™. Data processing steps included geometry installation, trace editing, timing corrections, elevation statics, muting, automatic gain control (AGC), bandpass filtering, velocity analysis, NMO correction, stacking, depth conversion, and deconvolution. The images were migrated by using Kirchhoff pre-stack depth migration (Schneider, 1978) to properly image complex structures, such as dipping layers and abruptly terminated layers. Finally, a high frequency post-stack bandpass filter (100-200-400-800 Hz) was applied to the images to help identify faults and formation contacts.

Seismic Data

The range of frequencies that could be used to measure first-arrival refractions was limited because there were high levels of cultural noise caused by refractions from urban features such as underground pipes and cement sidewalks, which made it difficult to determine the precise timing of the first-arrival refractions from the sedimentary strata. To minimize the effect of refractions from urban features, data were filtered using a relatively low-frequency bandpass filter (10-20-40-80 Hz). As a result, accurate P-wave seismic velocity images only could be developed for relatively shallow (~150 ft) depths along the seismic profiles by using relatively low-frequency (10 Hz minimum) data (fig. 12), which made it difficult to directly image small-scale (~30 to 50 ft) geological structures.

Seismic reflection images were developed for each of the seismic lines by stacking shot gathers with approximately 120 channels per shot and converting the time sections to depth sections by using both the inverted tomographic velocity and depth-extended velocity models developed for each line. Common Depth Point (CDP) intervals along Line 1 are 8 ft; hence, a seismic trace was produced every 8 ft along the lines. Several prominent reflectors were observed in the shot gathers within the first few hundred milliseconds, suggesting that the seismic energy penetrated more than 1,000 ft in depth. High frequency data (up to 200 Hz) could be used to develop seismic reflection images of the subsurface along the seismic profiles because (1) the presence of urban features did not affect high-frequency, secondary reflected, energy as much as first-arrival refracted energy, and (2) the first-arrival refractions were muted before stacking the reflection data.

The ability to use high-frequency data made it possible to image relatively small-scale features. The resolving ability of reflected seismic energy is about one-quarter of a wavelength (Dobrin and Savit, 1988). In theory, for the shallow subsurface with velocities of approximately 2,600 feet per second (ft/s), subsurface structures only a few feet or less in size could be imaged with the reflection data.

Seismic Velocity Images

Average seismic velocities (averaged over 100 ft or more) were measured along each seismic line from the surface to a maximum depth of about 150 ft bls (fig. 12). Except near the centers of Lines 1 and 2 (near Agua Caliente Spring), the near-surface velocities ranged from about 1,300 to about 2,300 ft/s (~400 to 700 meters per second [m/s]) in the upper 65 ft. From about 65 ft to about 150 ft depth, the velocities were highly variable, ranging from about 2,600 to 3,900 ft/s (~800 to 1,200 m/s), with higher localized velocities.

Average seismic velocities along Line 1 varied from about 1,800 to 3,800 ft/s (~550 m/s to 1,150 m/s), with near-surface (upper 50 ft) velocities along most of the seismic line ranging from about 1,800 to 2,300 ft/s (~550 to about 700 m/s) (fig. 12A). However, between lateral distances of about 1,000 to 1,300 ft along Line 1, average velocities as high as 2,625 ft/s (~800 m/s) were observed in the upper 30 ft. This relatively high-velocity zone coincides with a known perched water table (figs. 5 and 12A). Sediments below the relatively high-velocity layer (perched water table) probably are lower in velocity, causing a low-velocity or shadow zone. Such an inverted velocity structure prevents refracted arrivals from returning to the surface, making it impossible to measure the velocities of the underlying localized lower velocity layers with surface-based refraction methods. The measured P-wave velocities in the area of the known perched water table were less than 4,920 ft/s (1,500 m/s), which is the minimum velocity expected for saturated sediments (Schon, 1996, Catchings and others, 1999; 2008; 2009). This apparent contradiction is the result of using relatively low frequency (~10 to 30 Hz) seismic data to develop the velocity images and the limited thickness of the perched water table (about 30 ft; fig. 6). The low frequency seismic data developed for this study could not image the 30-ft perched water table as a discrete feature. The measured velocities are an average of the velocity of the high-velocity saturated sediments within the perched water table and low-velocity sediments above and below the perched water table.

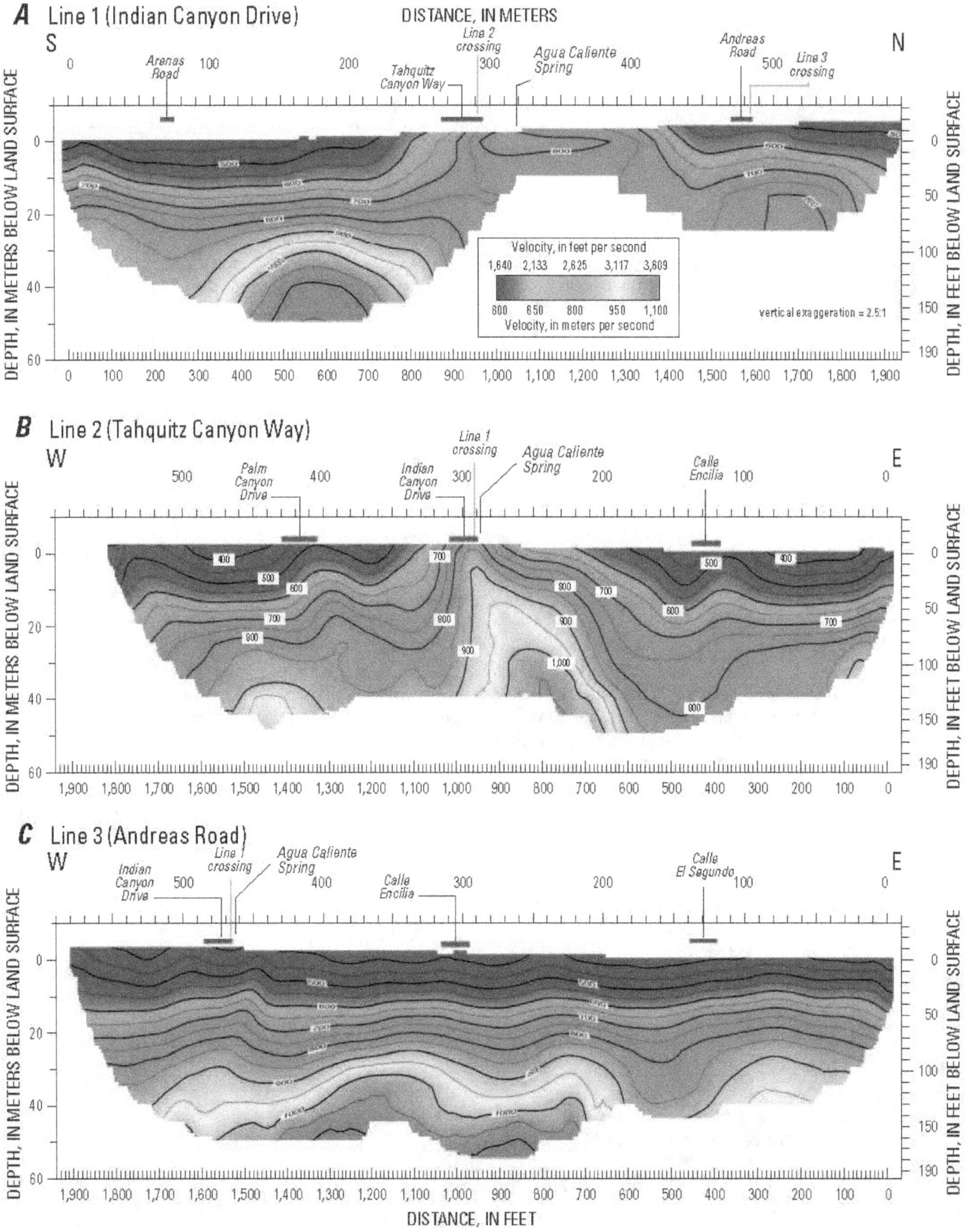

Figure 12.   P-wave seismic velocity images along (*A*) Line 1, (*B*) Line 2, and (*C*) Line 3 in the Agua Caliente Spring study area, California.

Average velocities along Line 2 range from less than about 1,300 ft/s (400 m/s) near the surface to about 3,600 ft/s (1,100 m/s) at depth near the center of the line (fig. 12B). A relatively high-average-velocity body (2,625 ft/s [800 m/s]) was imaged in the upper 30 ft near Agua Caliente Spring, in the center of Line 2 (figs. 11 and 12B). Relatively low-average velocities (1,300 to 2,300 ft/s [400 to 700 m/s]) were imaged in the upper 30 ft across the remainder of the seismic profile. The zone of higher average velocities coincides with the similar velocities observed along Line 1 where the two lines intersect, and this relatively high-velocity zone coincides with the perched water table surrounding the Agua Caliente Spring (figs. 5, 12A, and 12B). The high-velocity body is located east of Indian Canyon Drive and north of Tahquitz Canyon Way. The western flank of the high-velocity body terminates abruptly at Indian Canyon Drive. Drains beneath Indian Canyon Drive probably limit the western extent of the perched water table (fig. 5).

Average velocities along Line 3 range from less than about 1,300 ft/s (400 m/s) near the surface to about 3,600 ft/s (1,100 m/s) at about 150 ft bls (fig. 12C). Relatively low average velocities (about 1,300 to 2,300 ft/s [about 400 to 700 ms]) extend along the entire profile in the upper 50 ft, suggesting that the perched water table is not present beneath Line 3.

Reflection Images

Unmigrated Stacked Reflection Images

The unmigrated stacked reflection images are highly diffractive, particularly at depths in excess of 300 ft bls (fig. 13); therefore, these images primarily were used to evaluate the near-surface strata (upper 300 ft). The unmigrated stacked reflection image of Line 1 shows that the near-surface strata generally are layered and subhorizontal (fig. 13A). The near-surface strata appear to be disrupted beneath Tahquitz Canyon Way and Andreas Road, but these disruptions probably are an artifact of diffracted energy associated with the roads and urban features, such as drains, beneath the roads. The unmigrated stacked reflection image of Line 2 also shows that the near-surface strata largely are layered and subhorizontal with few disruptions (fig. 13B). The near-surface strata west of Calle Encilia appear to dip from east to west, but this may be an artifact of diffracted energy from the road and associated urban features beneath the road, such as drains. The unmigrated stacked reflection image of Line 3 shows that the near-surface strata are not well layered above about 150 ft bls; however, there is a prominent reflector at about 150 ft bls that extends along the entire length of the profile (fig. 13C). Below

the prominent reflector at about 150 ft bls, there appear to be sub-horizontally layered strata to depths of about 300 ft, but the stacked image is highly diffractive, making it difficult to correlate the reflectors laterally. No clear basement reflectors were observed in any of the unmigrated stacked reflection images along Lines 1–3 (figs. 13A–C).

Migrated Seismic Reflection Images

Because the unmigrated stacked images were highly diffractive, especially at depths in excess of 300 ft bls, pre-stack depth migration (PSDM) was used to better image the deeper strata (figs. 14A, 15A, and 16A). A high frequency post-stack bandpass filter (100-200-400-800 Hz) was applied to the images to improve the resolution of geologic structure and stratigraphic contacts. The seismic velocity image and gravity calculated depth to basement complex was added to the PSDM images to assist in the interpretation of the data (figs. 14B, 15B, and 16B).

The PSDM image of Line 1 shows reflections to about 1,100 ft bls on the southern end of the profile and to about 700 ft bls on the northern end of the profile (fig. 14A). The depth to the top of the basement complex likely corresponds to the base of the reflections, which generally coincides with the gravity-calculated depth to basement complex (fig. 14B). The seismic interpreted depth to basement complex is about 830 ft bls beneath the Agua Caliente Spring (fig. 14B). The PSDM image shows a change in reflective character of the strata above the seismic interpreted depth to basement complex. The strata above this contact are characterized by multiple prominent reflectors, whereas the strata below the contact are characterized by relatively weak reflectors. This change in reflective character is interpreted as the contact between unconsolidated valley-fill deposits and indurated (partly consolidated to consolidated) valley-fill deposits. The indurated valley-fill deposits are denser than the unconsolidated valley-fill deposits; therefore, much of the seismic energy reflects off of this boundary, reducing that available to reflect off of deeper strata. Unfortunately, refraction velocities were not available to the depths of the interpreted contact to confirm this interpretation (fig. 14B). Similar to the interpreted depth to basement complex, the interpreted depth to the indurated valley-fill deposits decreases from south to north; reaching a minimum of about 200 ft bls beneath Calle Encilia (fig. 14B). The strata above about 200 ft bls are layered and subhorizontal with few interruptions. The strata beneath about 200 ft bls west of Tahquitz Canyon Way are less reflective and appear to dip from north to south. The change in reflective character and apparent dip of the strata suggests the presence of a fault.

A

Line 1 (Indian Canyon Drive)

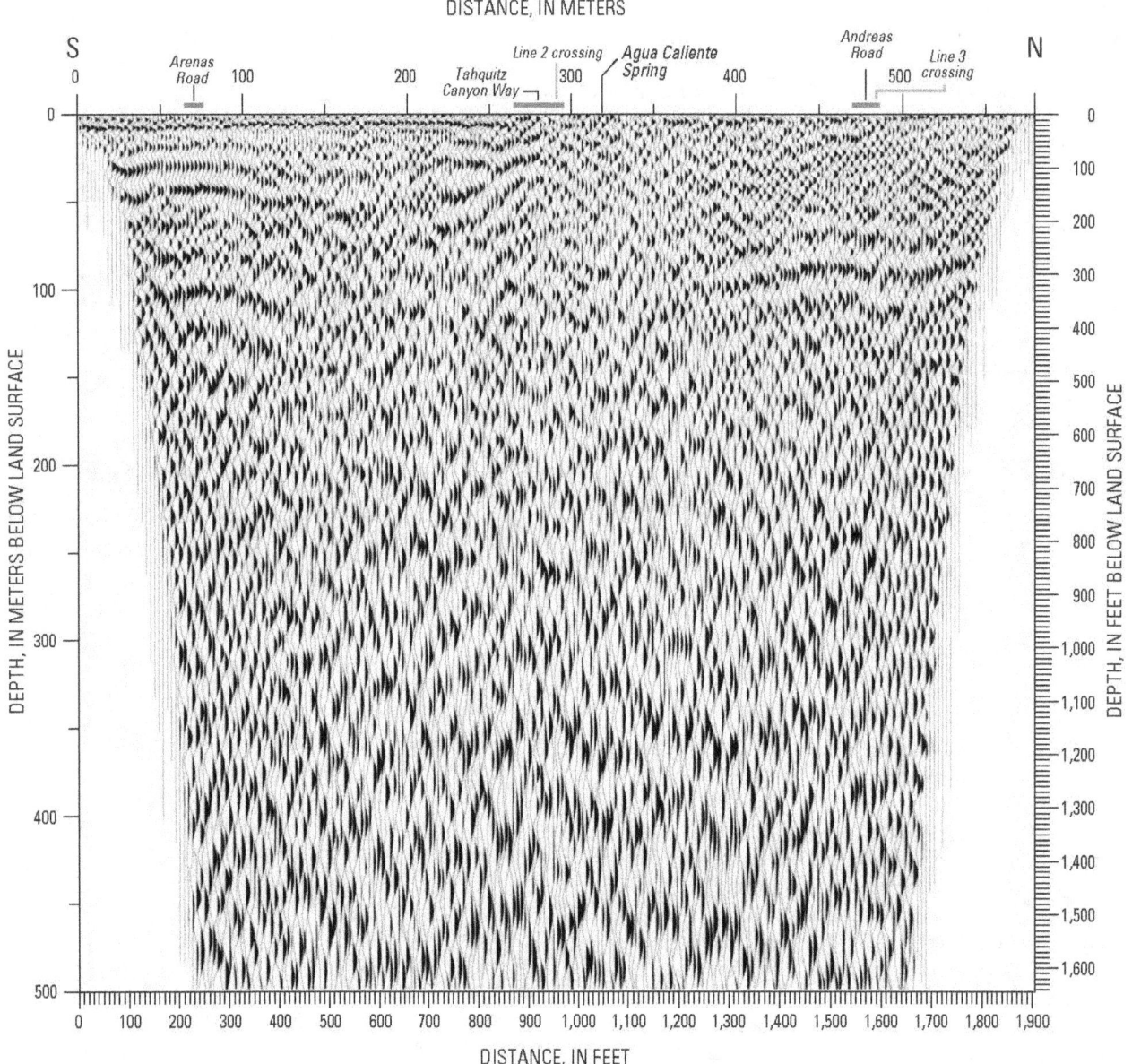

Figure 13.   Unmigrated stacked seismic reflection image of (*A*) Line 1, (*B*) Line 2, and (*C*) Line 3 of the Agua Caliente study area, California.

B

Line 2 (Tahquitz Canyon Way)

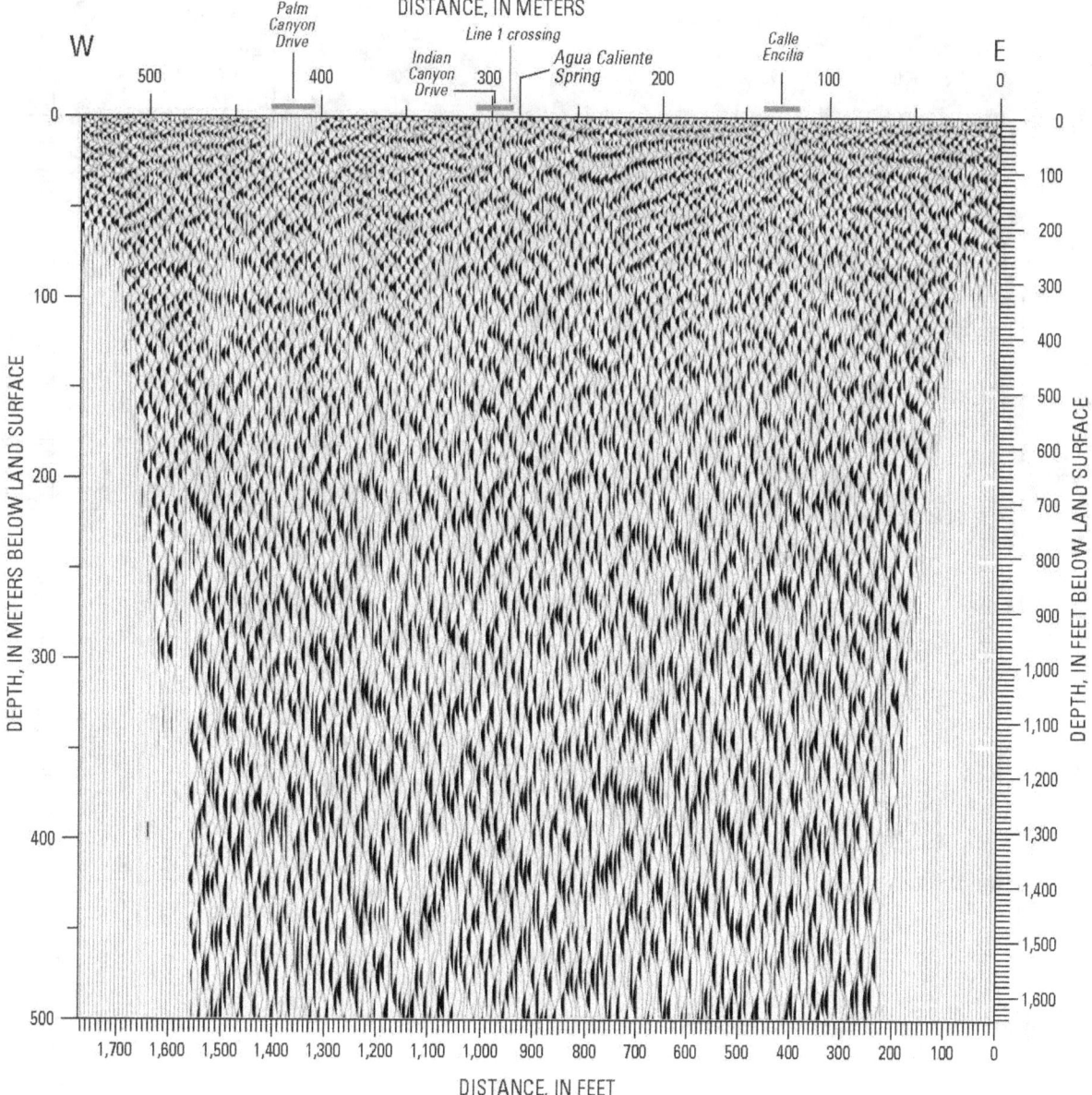

Figure 13.—Continued

C

Line 3 (Andreas)

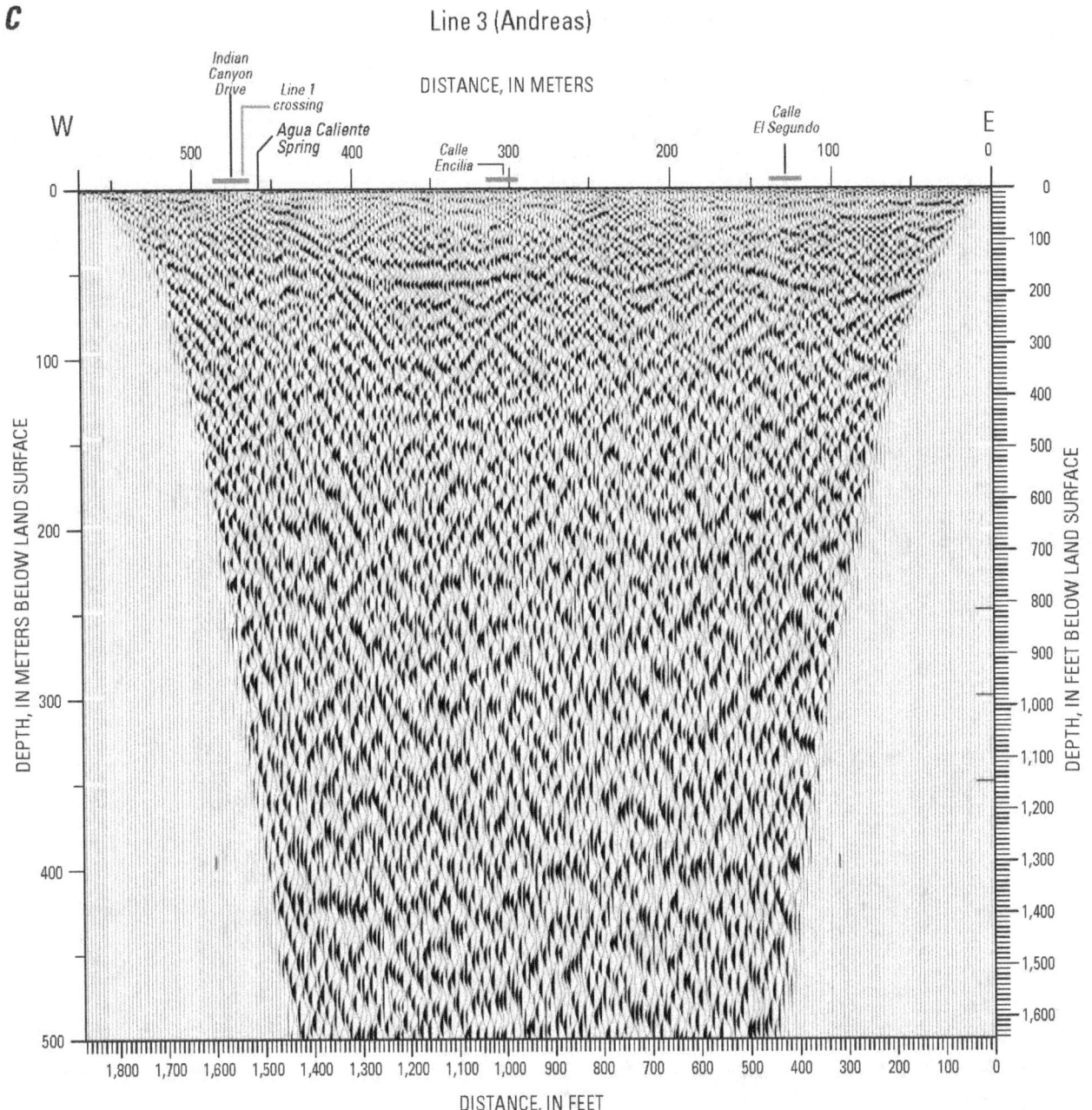

Figure 13.—Continued

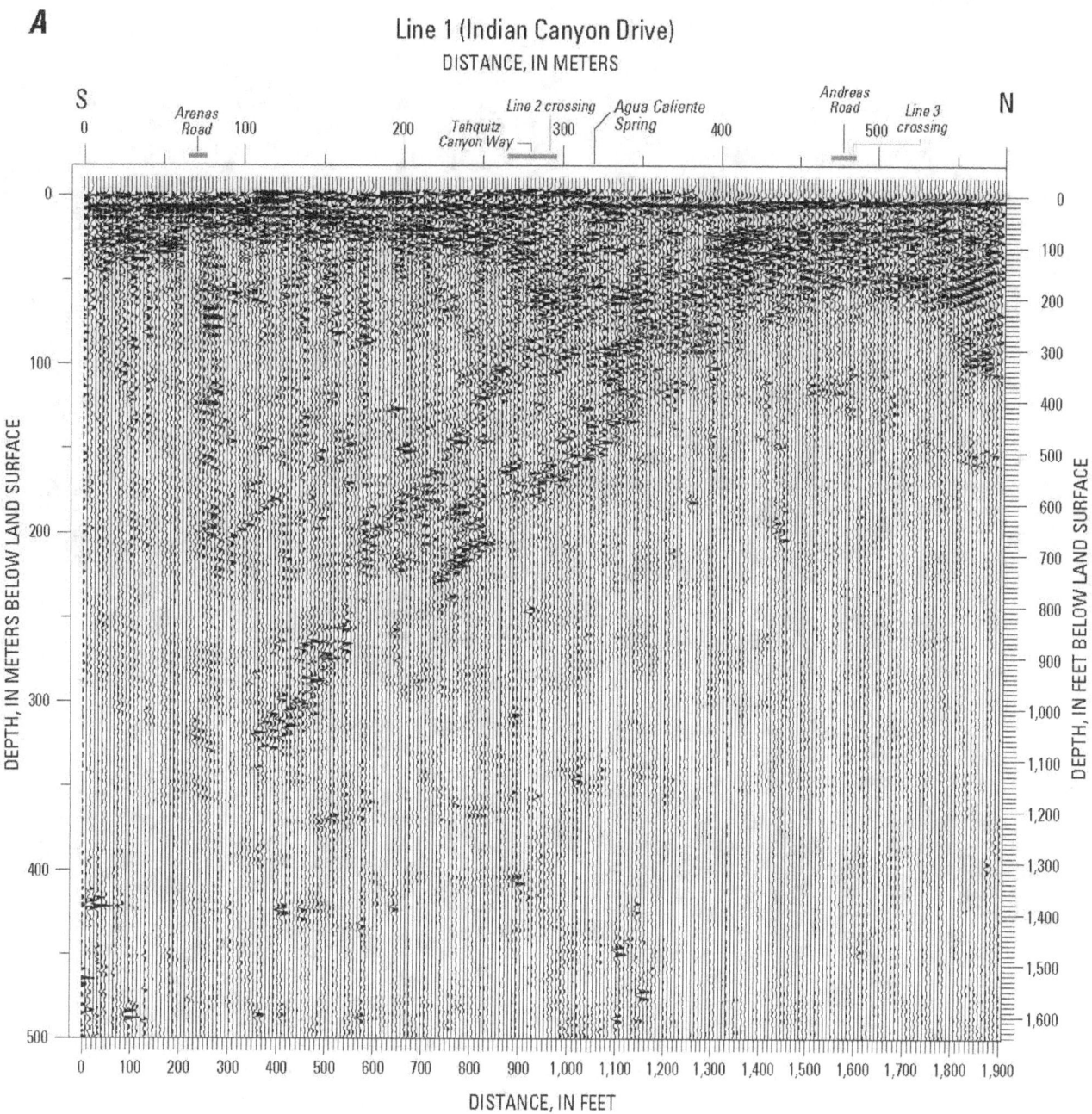

A

Line 1 (Indian Canyon Drive)
DISTANCE, IN METERS

Figure 14. Migrated seismic reflection image along (*A*) Line 1 with (*B*) superimposed P-wave seismic velocity image, gravity-calculated depth to basement complex, and interpreted faults and contacts in the Agua Caliente Spring study area, California.

B

Line 1 (Indian Canyon Drive)
DISTANCE, IN METERS

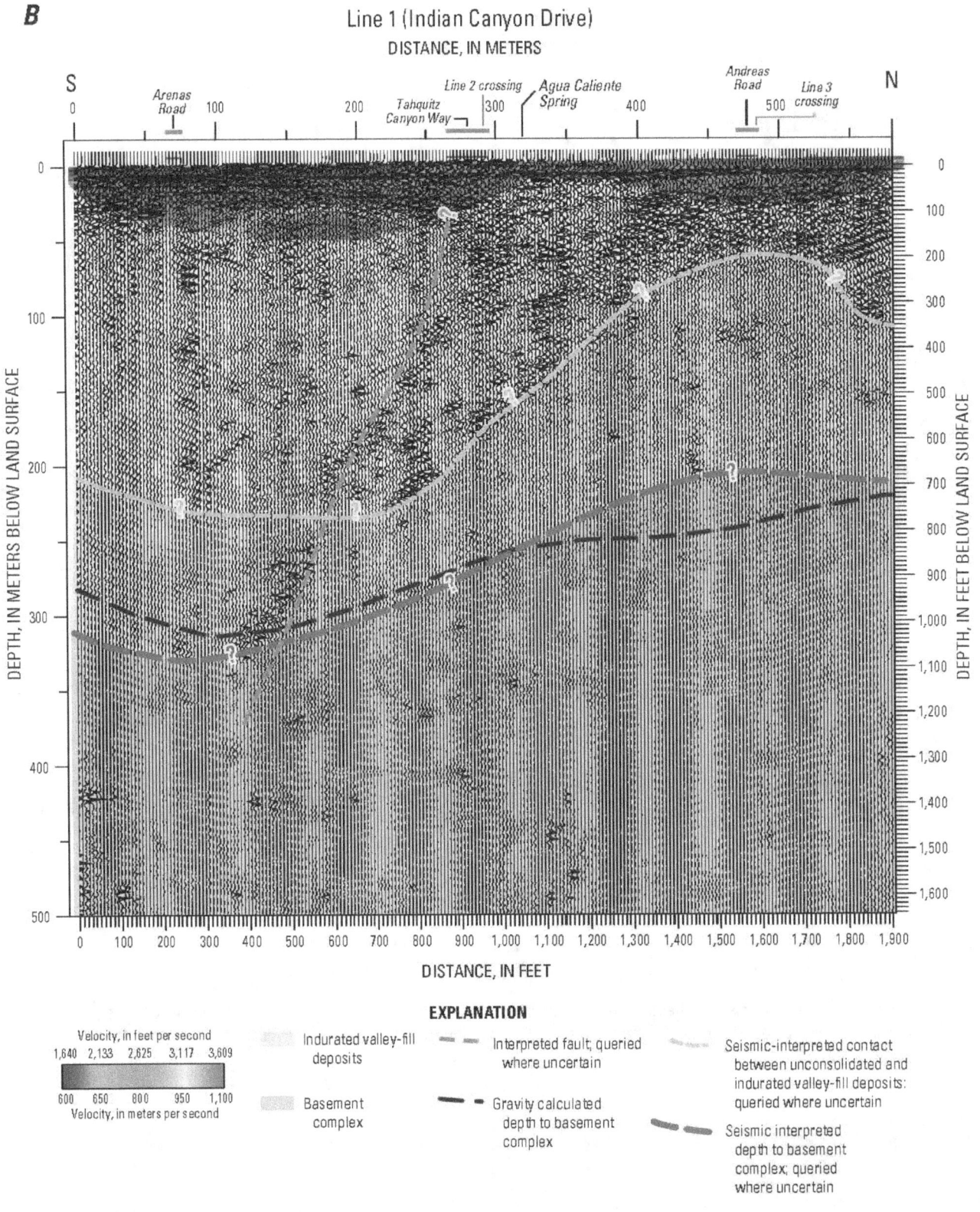

EXPLANATION

Velocity, in feet per second

| 1,640 | 2,133 | 2,625 | 3,117 | 3,609 |

| 600 | 650 | 800 | 950 | 1,100 |

Velocity, in meters per second

Indurated valley-fill deposits

Basement complex

Interpreted fault; queried where uncertain

Gravity calculated depth to basement complex

Seismic-interpreted contact between unconsolidated and indurated valley-fill deposits: queried where uncertain

Seismic interpreted depth to basement complex; queried where uncertain

Figure 14.—Continued

A

Line 2 (Tahquitz Canyon Way)

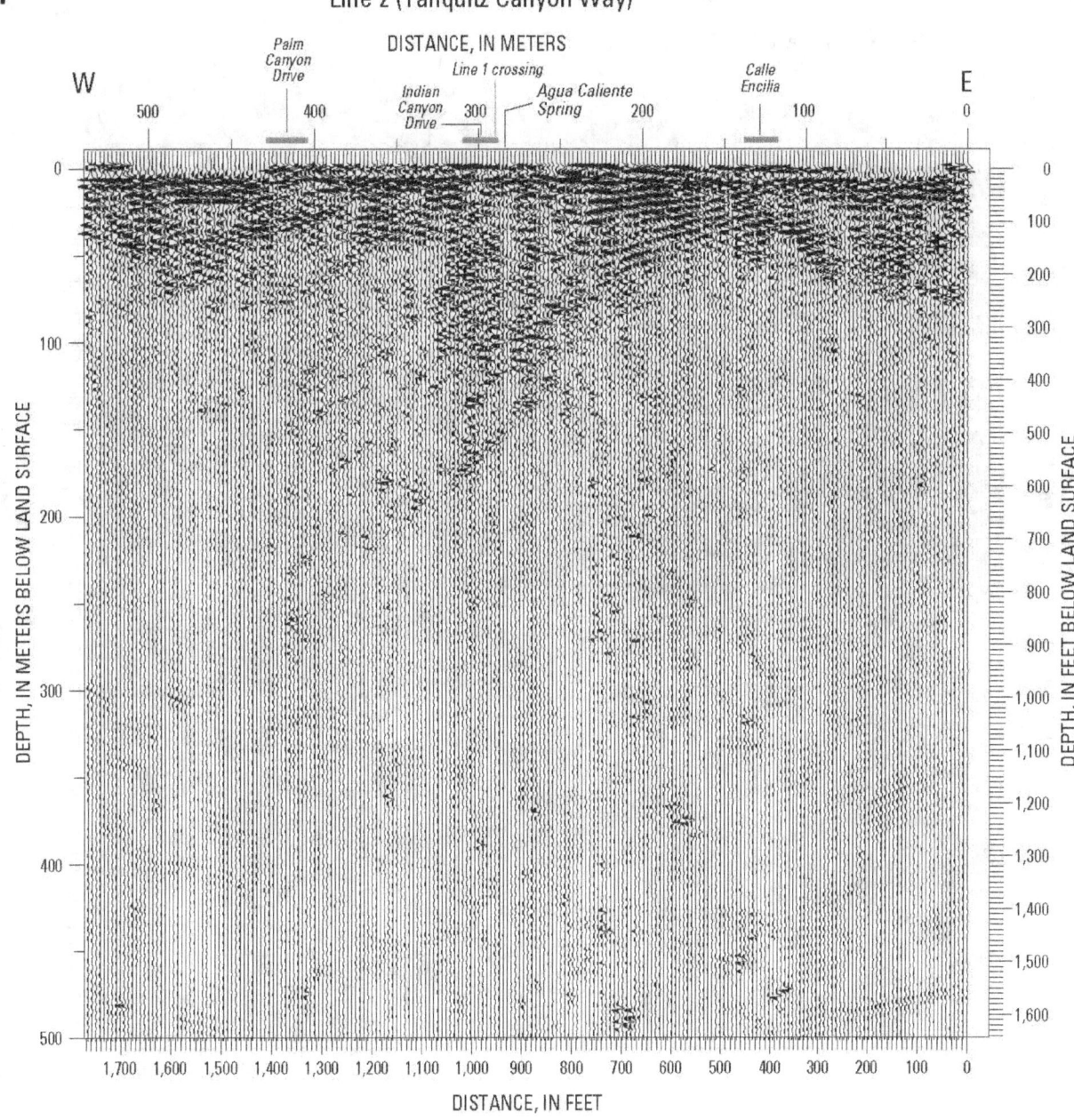

Figure 15. Migrated seismic reflection image along (*A*) Line 2 with (*B*) superimposed P-wave seismic velocity image, gravity-calculated depth to basement complex, and interpreted faults and contacts in the Agua Caliente Spring study area, California.

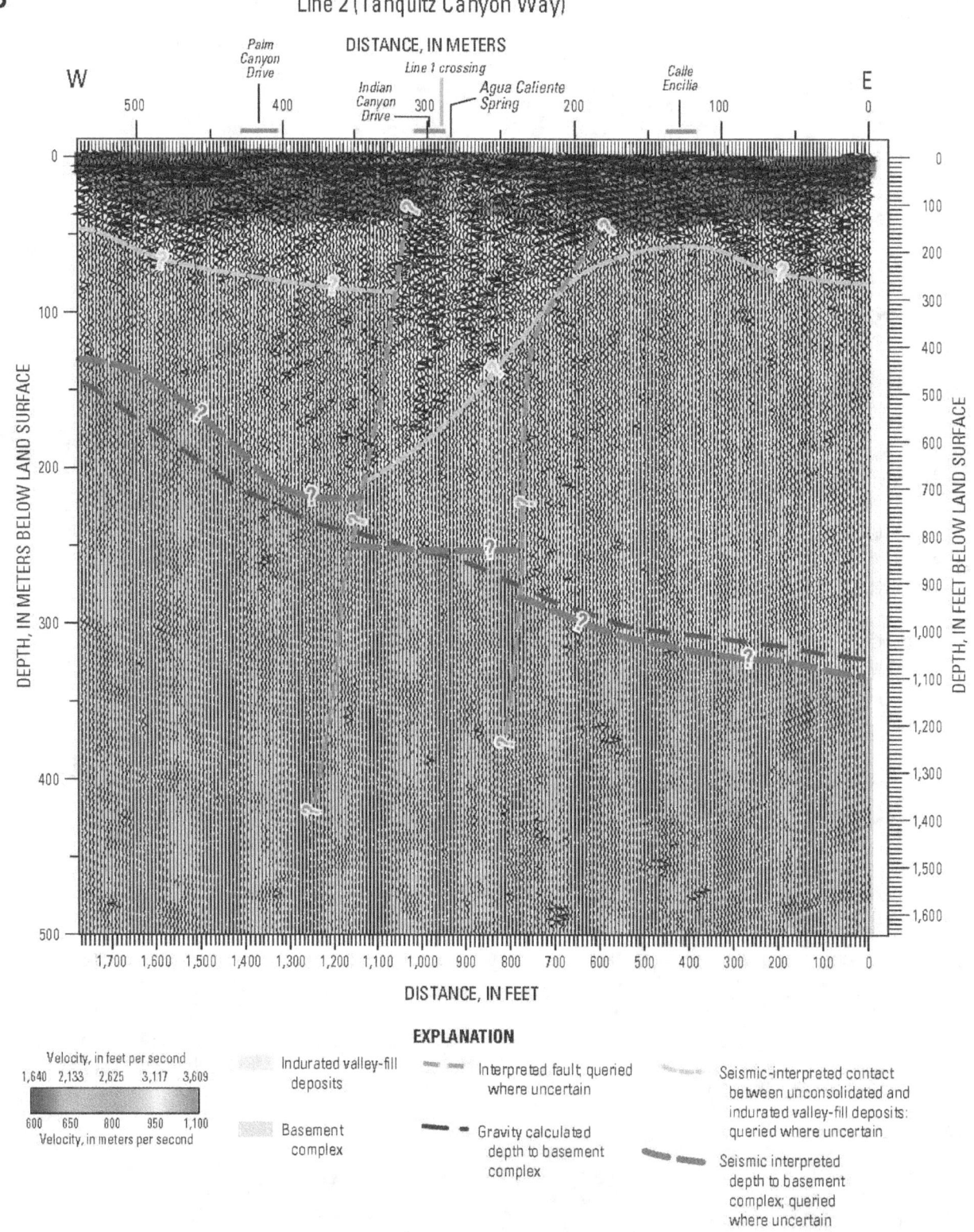

B

Line 2 (Tahquitz Canyon Way)

Figure 15.—Continued

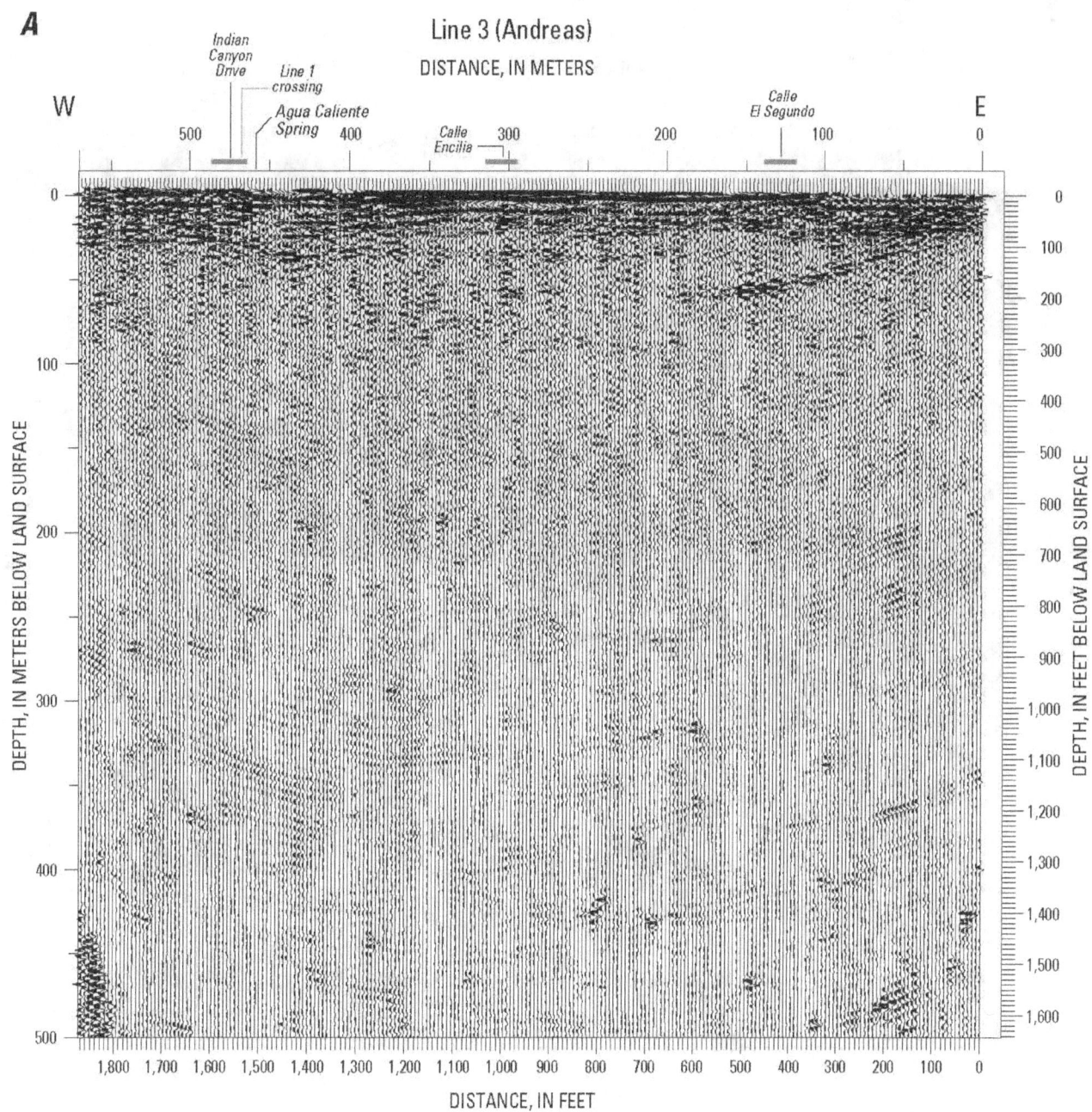

Figure 16.   Migrated seismic reflection image along (*A*) Line 3 with (*B*) superimposed P-wave seismic velocity image, gravity-calculated depth to basement complex, and interpreted faults and contacts in the Agua Caliente Spring study area, California.

B

Line 3 (Andreas)

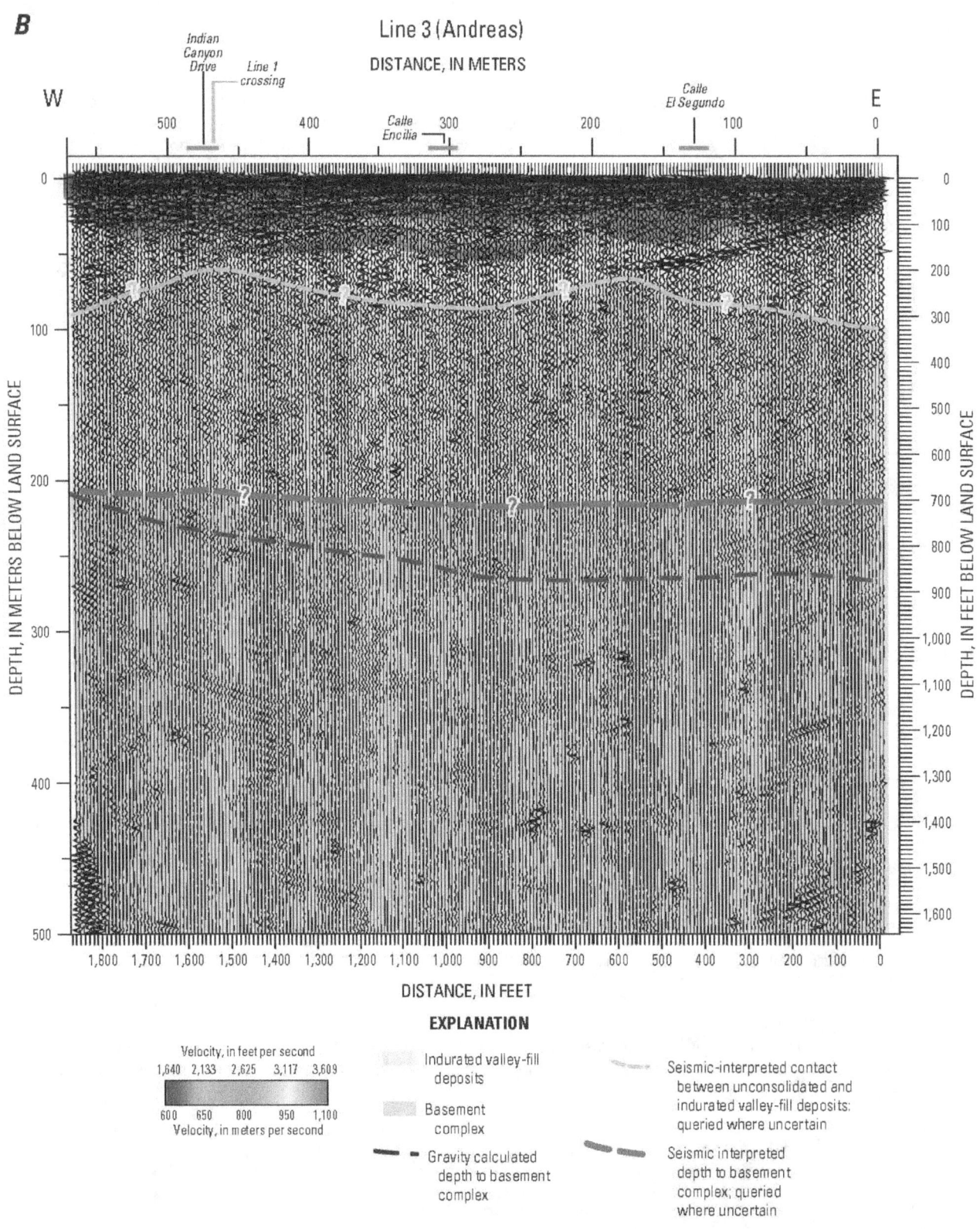

Figure 16.—Continued

The PSDM image of Line 2 shows reflections to about 500 ft bls on the western end of the profile and to more than 1,000 ft bls on the eastern end of the profile which are interpreted as the depth to the basement complex (fig. 15B). The seismic interpreted depth to basement complex is similar to gravity-calculated depth to basement complex; however, the depth of the basement complex is difficult to determine along the profile because of the lack of reflectors in the deeper strata (fig. 15B). Similar to line 1, most of the seismic energy is reflected at the interpreted contact between the unconsolidated valley-fill deposits and the indurated valley-fill deposits. This contact is at about 200 to 300 ft bls on the western end of the profile, drops to more than 500 ft bls beneath Indian Canyon Drive, and then rises rapidly east of Indian Canyon Drive to a high of about 200 ft bls beneath Andreas Road. The strata above about 200 ft bls are layered and subhorizontal with few interruptions. The displacement of the contact between the unconsolidated valley-fill deposits and the indurated valley-fill deposits west of Indian Canyon Drive suggests the presence of a fault. Disruptions in the layering and changes in the character of reflectors between Indian Canyon Drive and Calle Encilia suggest the presence of a second fault. These interpreted faults may be the continuation of the Palm Canyon fault inferred from the gravity data (figs. 9 and 15B).

The PSDM image of Line 3 shows reflections to about 700 ft bls along the entire profile (fig. 16A). The seismic interpreted depth to basement complex along Line 3 is similar to the gravity-calculated profile. Similar to lines 1 and 2, most of the seismic energy is reflected at the interpreted contact between the unconsolidated valley-fill deposits and the indurated valley-fill deposits. This contact is at about 200–300 ft bls beneath the entire profile, and the strata above the contact are layered and subhorizontal with no major disruptions. The faults interpreted along lines 1 and 2 were not apparent along line 3, suggesting that the north-south trending Palm Canyon fault steps to the west near the Agua Caliente Spring, as indicated by the gravity data (fig. 9A). The gravity-interpreted location of the Palm Canyon fault lies to the west of the western end of line 3 (figs. 9 and 11).

Summary of Seismic Imaging Results

Shallow-depth seismic refraction and reflection surveys were conducted along three lines near the Agua Caliente Spring to help define the stratigraphy and geologic structures associated with the spring. The seismic imaging results along the three lines are presented in a fence diagram to help show stratigraphic interrelation and structure in the Agua Caliente Spring study area (fig. 17).

Relatively high cultural noise levels in downtown Palm Springs did not allow the collection of seismic velocity images below about 150 ft bls. Consistent with observations from nearby wells, the seismic velocity images suggest that a perched water table occurs in the upper 30 ft of sediments, north and east of the intersection of Indian Canyon Drive and Tahquitz Canyon Way (fig. 12). The seismic velocity images indicate that the perched water is present only near the spring, suggesting that discharge from the spring is the source of the perched groundwater.

The seismic reflection data indicate that the depth to the basement complex is about 830 ft bls directly beneath the Agua Caliente Spring and that the depth to the basement complex decreases from south to north, indicating the presence of a buried basement ridge to the north of the Agua Caliente Spring, which is similar to what is shown by the gravity data. The migrated seismic reflection images indicate the presence of a density contrast above the seismic interpreted depth to basement complex, which is interpreted as the contact between overlying unconsolidated valley-fill deposits and underlying indurated valley-fill deposits. The seismic interpreted contact between the unconsolidated valley-fill deposits and the indurated valley-fill deposits is about 500 ft bls directly beneath Agua Caliente Spring and rises to about 200 ft bls less than 500 ft east and north of the spring (figs. 14B and 15B). The migrated seismic reflection images show disruptions in the layering and changes in the character of reflectors in the strata beneath the spring, suggesting the presence of two faults (figs. 15A and B), which probably are related to the north-south trending Palm Canyon fault (fig. 4). The faults were not identified along line 3, to the north of the Agua Caliente Spring, suggesting that the north-south trending Palm Canyon fault steps to the west near the spring, which also was inferred from the gravity data. Faults in consolidated rock, such as the basement complex and the indurated valley-fill deposits, can develop zones of sheared and broken rock that may be highly permeable (Freeze and Cherry, 1979); whereas, faults in unconsolidated deposits can form barriers to groundwater flow (Londquist and Martin, 1991). Faulting along the buried ridge of the basement complex and indurated valley-fill deposits could provide a pathway into overlying valley-fill deposits from an underlying reservoir of deep geothermal water, which is the probable source of the Agua Caliente Spring (fig. 10). If the fault zones are permeable in the seismic interpreted indurated valley-fill deposits, geothermal water could rise to within about 200 ft of land surface less than 500 ft from the Agua Caliente Spring.

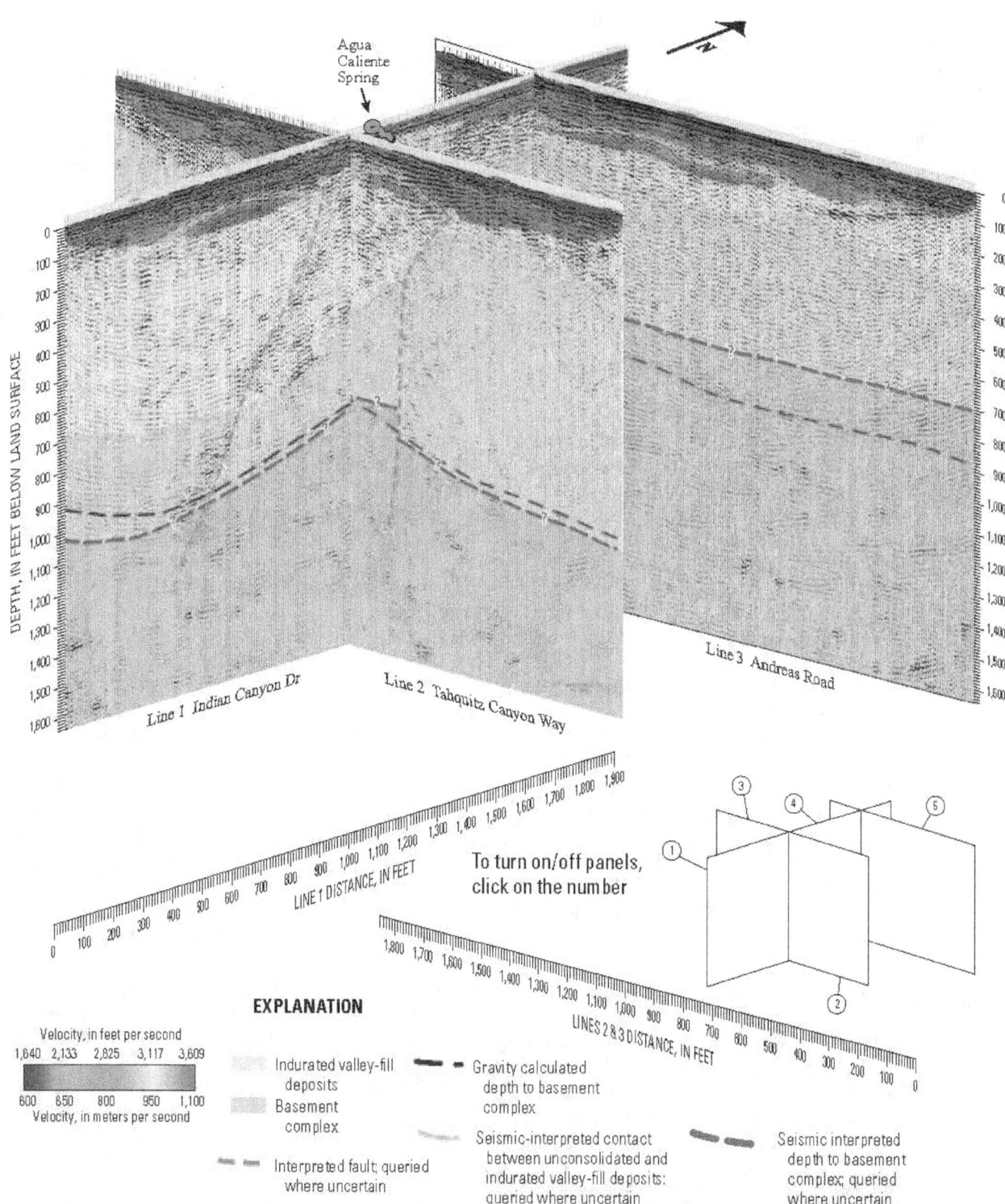

Figure 17. Seismic profiles showing interpreted depth to basement complex in the Agua Caliente Spring study area, California. Different panels of the fence diagram can be added or removed from view by clicking on the number of the panel in a schematic of the fence diagram.

Interferometric Synthetic Aperture Radar Data in the Agua Caliente Spring Area

By Michelle Sneed and Justin T. Brandt

Interferometric synthetic aperture radar (InSAR) was used in this study to help identify ground-surface deformation and locate structures, such as faults, that may affect groundwater movement. InSAR is a satellite-based remote sensing technique that can detect centimeter-level ground-surface deformation over a 38.6 mi^2 (100 square kilometers [km^2]) area with a spatial resolution of 295 ft (90 m) or less (Galloway and others, 2000). Data from the European Space Agency's (ESA) ENVISAT satellite were obtained for analysis. The InSAR technique uses two synthetic aperture radar (SAR) images of the same area acquired at different times and "interferes" (differences) them, resulting in maps called interferograms that show line-of-sight vertical ground-surface displacement (range change) between the two time periods. A complete description of the technique is available in Galloway and others (1999, 2000) and Sneed and Brandt (2007).

InSAR Calculations of Land-Surface Deformation

Eighteen interferograms representing time periods ranging from 35 to 595 days between October 26, 2003, and September 25, 2005, were analyzed to determine if vertical changes of the land surface have occurred in the study area. Results of the analysis indicate that little deformation has occurred in the study area for the time periods represented by the interferograms. The largest amount of subsidence shown on any of the interferograms was about 0.6 in. (15 millimeters [mm]), which is close to the resolution for this technique, in a small area about one mile south of the Agua Caliente Spring between January 23, 2005, and July 17, 2005, (fig.18). A large area surrounding the spring subsided about 0.4 in. (10 mm) during this same period (fig. 18). In this area, eight interferograms showed subsidence of 0.4 in. (10 mm) or less, three interferograms showed as much as 0.2 in (5 mm) of uplift, and seven interferograms showed no change. The InSAR data analyzed for this study indicate that little, if any, land subsidence occurred in the Agua Caliente Spring study area over this period.

InSAR Inferred Structure

Measurements of land-surface subsidence can be used to infer the location of buried faults not readily evident in surface expression (Galloway and others, 1999). In alluvial basins, some faults are barriers to groundwater flow; therefore, water-level changes and related land-surface deformation are greater on the side of the fault where pumping occurs. Although this area showed minimal deformation during the time periods represented by the interferograms, a linear deformation boundary is apparent about 1 mi west of the Palm Springs International Airport (fig. 18) in three interferograms that indicate measureable subsidence. The area west of the boundary shows subsidence of greater than 0.4 in. (10 mm), whereas the area east of the boundary shows essentially no subsidence. The deformation boundary trends northwest-southeast for about 2.5 miles, and lies about 1 mi west of the location of a northwest-southeast fault inferred by Jennings (1994) (fig. 4). None of the interferograms provide information on the location of possible buried faults near the Agua Caliente Spring.

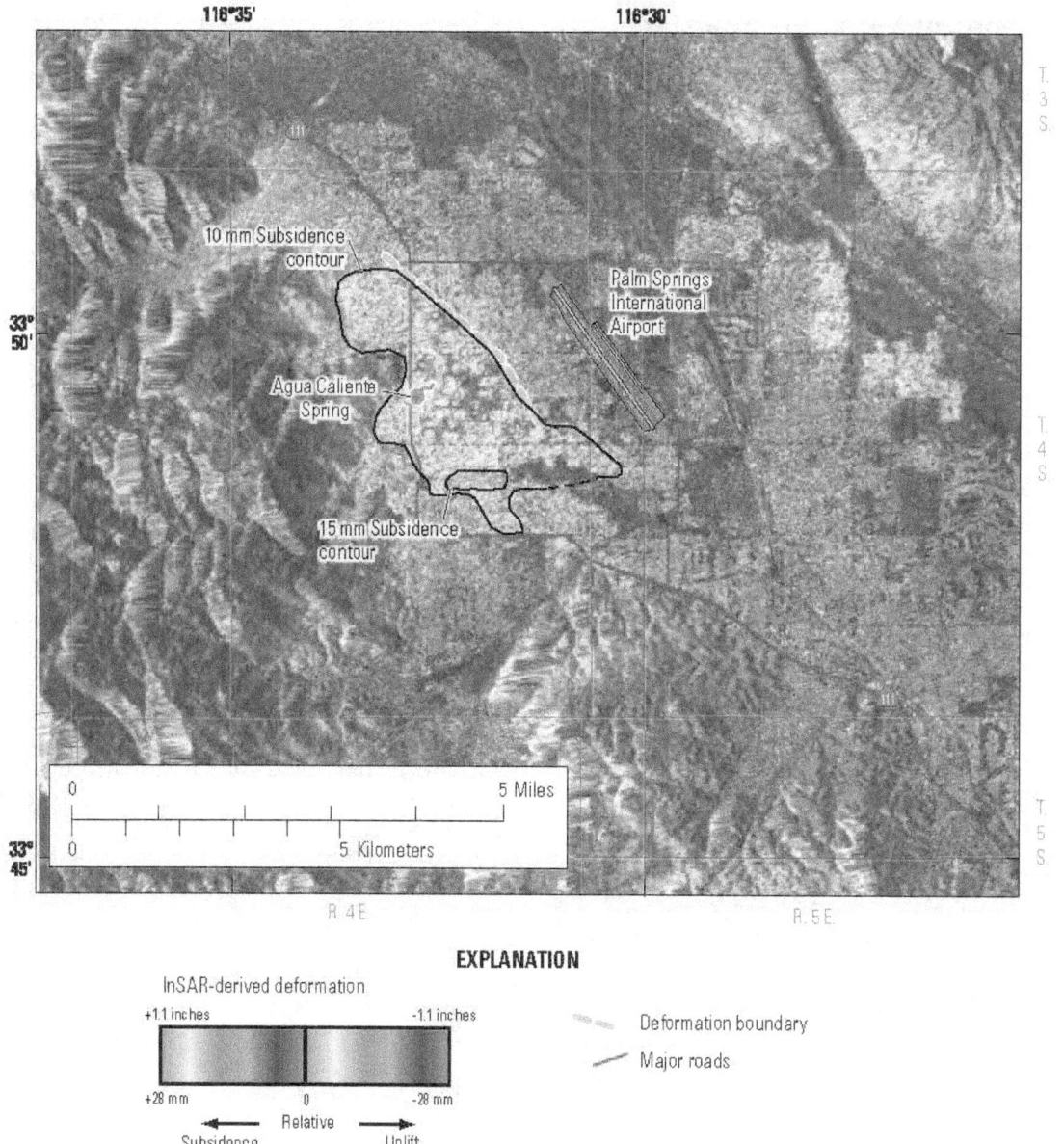

Figure 18.   Vertical changes in land surface from January 23, 2005, through July 17, 2005, and the location of a deformation boundary in the Agua Caliente Spring study area, California.

Spring Discharge and Temperature

By Gregory A. Smith, Roy A. Schroeder, and Keith J. Halford

Historical spring discharge and temperature measurements, and seasonal measurements collected for this study, were evaluated to determine if there is a relation between spring discharge and spring temperature. Seasonal discharge and temperature measurements also were collected from Fenced, Chino Warm, and Chino Cold Springs for this study.

Spring Discharge

Agua Caliente Spring

Historical measurements indicate that water discharge at the Agua Caliente Spring has ranged from 5 to 60 gal/min during the past century (table 4). Possible explanations for the historical variation include differences in methods used to measure discharge, fluctuations in climate, and(or) changes of the spring discharge area over time—from an uncontrolled discharge into a pond in the early 1900s to a controlled discharge into a steel collector tank since 1958. Uncertainties in the accuracy of historical measurements make it difficult to interpret the historical variations in spring discharge, so this study endeavored to obtain a more reliable record of discharge during its 2-year period of study. Discharge at Agua Caliente Spring was measured using two methods: (1) total-flow data and (2) water-level recovery curves. A detailed description of the second method can be found in Halford and Kuniansky (2002).

Total-Flow Data Method

Total flow was recorded by using flow meters previously installed at the two pumping outlets (spa pump and overflow pump) from the Agua Caliente Spring by personnel from the Agua Caliente Band of Cahuilla Indians. Flow-meter readings were obtained from July 2005 to February 2006 and from mid-July 2006 to September 2006. The spring's flow rate (discharge) is calculated by summing total cumulative flow from meter readings (in gallons [gal]) on the two pumps, subtracting this total from the previous recorded total, and then dividing this difference by the number of days elapsed between readings. Because the date, but not the exact time of day, was recorded, accurate flow rates can only be determined over reasonably long periods of time. Over periods of a month, the error introduced by recording only the day of measurement becomes negligible for purposes of discerning temporal or seasonal trends. Discharge from Agua Caliente Spring using this method averaged 19 gal/min over the entire period of record, 22 gal/min between July 2005 and February 2006, and 14 gal/min between mid-July 2006 and September 2006.

Water-Level Recovery Method

The Agua Caliente Band of Cahuilla Indians operates a pressure transducer at the Agua Caliente Spring to record water levels in the steel collector tank. Transducer data are available for April through September 2004, November 2004 through January 2005, August through September 2005, and June through August 2006. Spring water is pumped from the steel collector tank for use at the Spa Hotel, and the pressure transducer records the water-level response to these pumping events. Water-level recovery curves from this record were used to estimate the spring's discharge. The slope of the water-level recovery curve (in feet per minute [ft/min]) was multiplied by the known volume of the tank per vertical foot (in cubic feet per foot [ft^3/ft]) to estimate the discharge rate (in cubic feet per minute [ft^3/min]). Slopes were calculated for six pumping events on the 14th day of each month and then averaged to compute monthly discharge rates (fig. 19). Discharge estimated using this water-level recovery method averaged 16.8 gal/min between April 2004 and January 2005, 23.9 gal/min between August and September 2005, and 8.9 gal/min between June and August 2006 (fig. 20).

Relation Between Spring Discharge and Precipitation

Discharge measured and estimated at the Agua Caliente Spring during this study was compared to simulated annual precipitation to determine if they are positively correlated (fig. 21A). Note that simulated annual precipitation is presented on a calendar-year (January 1–December 31) and water-year (October 1–September 30) basis. The calendar year does not quite match the sampling design for this study because it incorporates precipitation from the latter part of one wet season (January–March) and the initial part of the following wet season (October–December). For this reason, precipitation totals also are presented for water years, which represent precipitation from a single wet season (October–March).

Discharge was highest during the summer of 2005, followed by 2004, and was lowest during the summer of 2006. As indicated on figure 21A, annual calendar-year precipitation in the study area exceeded the mean in 2004 and 2005 (wetter) and was below the mean in 2006 (dryer). When water-year precipitation totals are compared, precipitation was 2.5 times higher during water year 2005 than water year 2006. These observations suggest that the discharge of Agua Caliente Spring is influenced by recent precipitation, although discharge needs to be measured over a period spanning multiple wet and dry cycles to establish the relationship with a high degree of confidence.

Table 4. Discharge, temperature, and water-quality data collected at the Agua Caliente Spring, California,1876–2006.

[See figure 1 for site location. **Abbreviations:** mm/dd/yyyy, month/day/year; gal/min, gallons per minute; °C, degrees Celsius; μS/cm, microsiemens per centimeter at 25 °C; mg/L, milligrams per liter; μg/L, micrograms per liter; E, value estimated; R, value reported; Trace, present but not quantified for historical analysis with high reporting level; <, less than; –, no data]

Date (mm/dd/yyyy)	Discharge (gal/min)	Temperature, water (°C)	pH (standard units)	Specific conductance (μS/cm)	Sodium (mg/L)	Potassium (mg/L)	Calcium (mg/L)	Magnesium (mg/L)	Boron (μg/L)	Chloride (mg/L)	Fluoride (mg/L)	Iron (μg/L)	Carbonate (mg/L)	Bicarbonate (mg/L)	Sulfate (mg/L)
[1]1876	–	–	–	–	158	–	Trace	Trace	–	188	–	–	47	–	Trace
[1]1908	5 E	–	–	–	–	–	–	–	–	–	–	–	–	–	–
[2]12/19/1917	10 E	37.8	–	–	–	–	1.7	0.39	–	23	–	0.24	40	28	39
[3]09/06/1938	10 E	–	–	–	41	–	13	Trace	0.2	14	–	–	Trace	103	19
[3]05/04/1905	60 R	–	–	–	–	–	–	–	–	–	–	–	–	–	–
[4]07/16/1953	21 R	–	–	–	–	–	–	–	–	–	–	–	–	–	–
[4]07/17/1953	23 R	–	–	–	–	–	–	–	–	–	–	–	–	–	–
[4]07/18/1953	24 R	–	–	–	–	–	–	–	–	–	–	–	–	–	–
[4]07/19/1953	23 R	–	–	–	–	–	–	–	–	–	–	–	–	–	–
[4]07/20/1953	25 R	–	–	–	–	–	–	–	–	–	–	–	–	–	–
[5]10/28/1953	–	42.2	9.50	323	68	1.3	4.2	2.5	0	20	2.4	–	21	51	–
[5]06/15/1958	–	40.6	9.00	332	70	0.4	0	0	0.11	24	2.5	–	24	46	–
[5]05/12/1959	25	41.7	–	–	–	–	–	–	–	–	–	–	–	–	–
[6]1981	–	41.0	9.10	310	56	3	3	1	0.1	26	2.2	–	34	32	–
01/28/2005	–	–	–	330	–	–	–	–	–	–	–	–	–	–	–
04/27/2005	–	41.2	9.80	323	67.5	0.6	1.3	<0.008	0.144	22.5	2.68	0.004	19	63	25.3
09/09/2005	–	41.0	9.70	342	65	0.59	1.35	<0.008	0.143	23.4	2.78	<0.006	20	63	28
04/05/2006	–	41.0	9.70	336	66.6	0.62	1.3	<0.004	0.136	22.8	2.77	0.005	20	51	28.2
09/11/2006	–	41.7	9.70	328	65	0.63	1.23	<0.004	0.146	22.3	2.64	<0.006	19	59	26.7

[1] Waring (1915).

[2] Brown (1923).

[3] Garrett and Dutcher (1951).

[4] Dutcher (1953).

[5] Dutcher and Bader (1963).

[6] Leivas and others (1981).

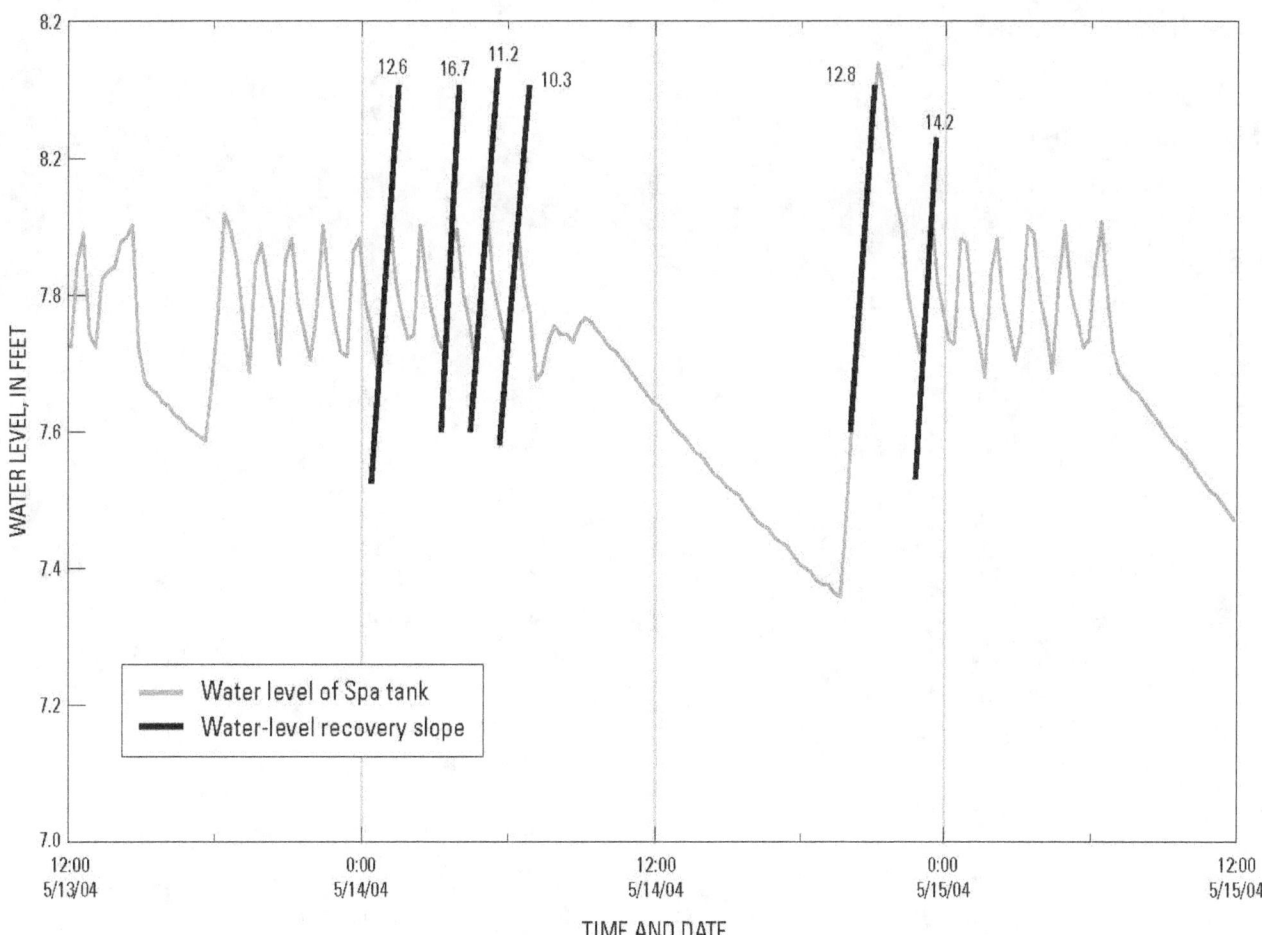

Figure 19. The water-level recovery method for estimating discharge from the Agua Caliente Spring, California, May 13–15, 2004.

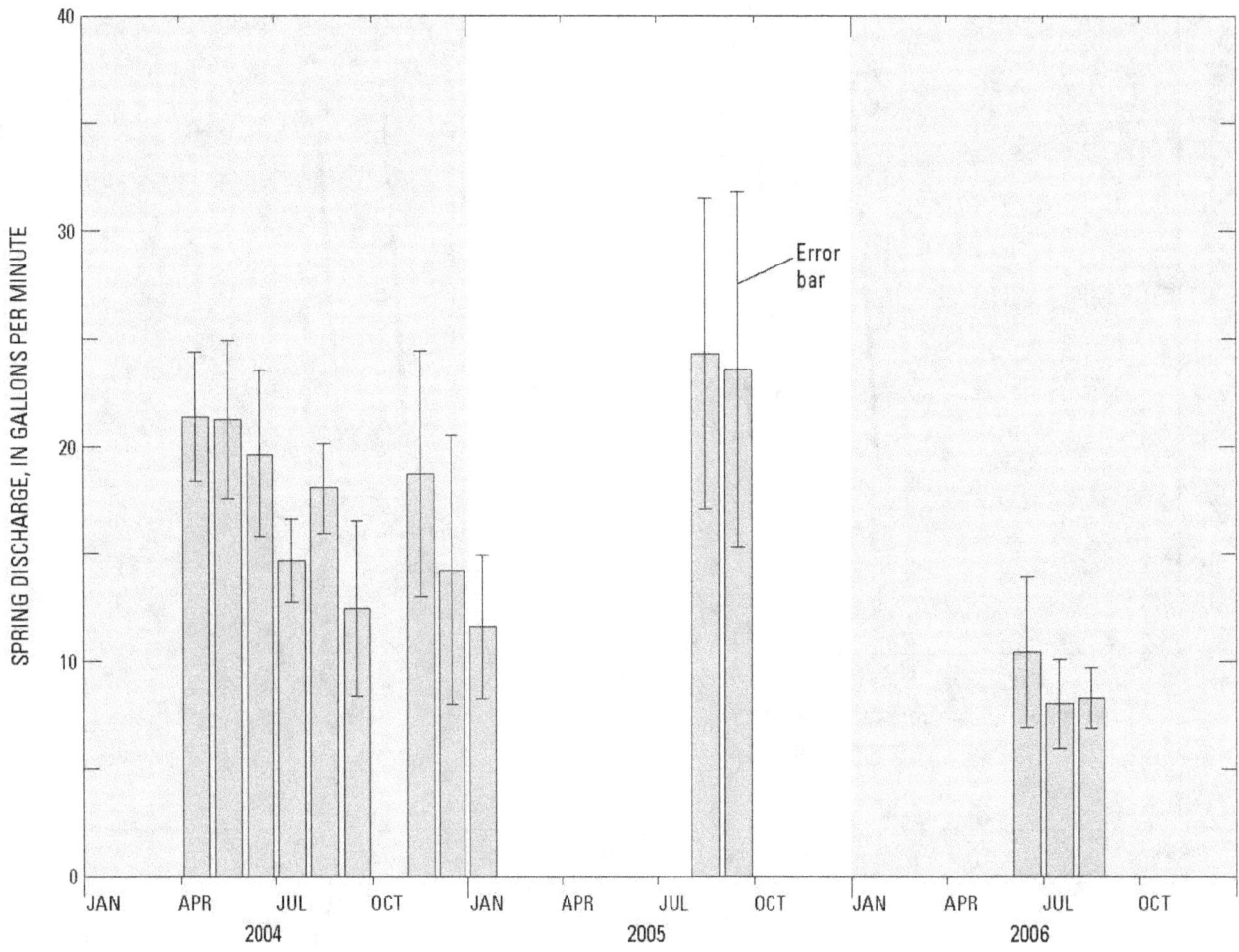

Figure 20.   Estimated discharge of the Agua Caliente Spring, California, using water-level recovery method, 2004–06.

A

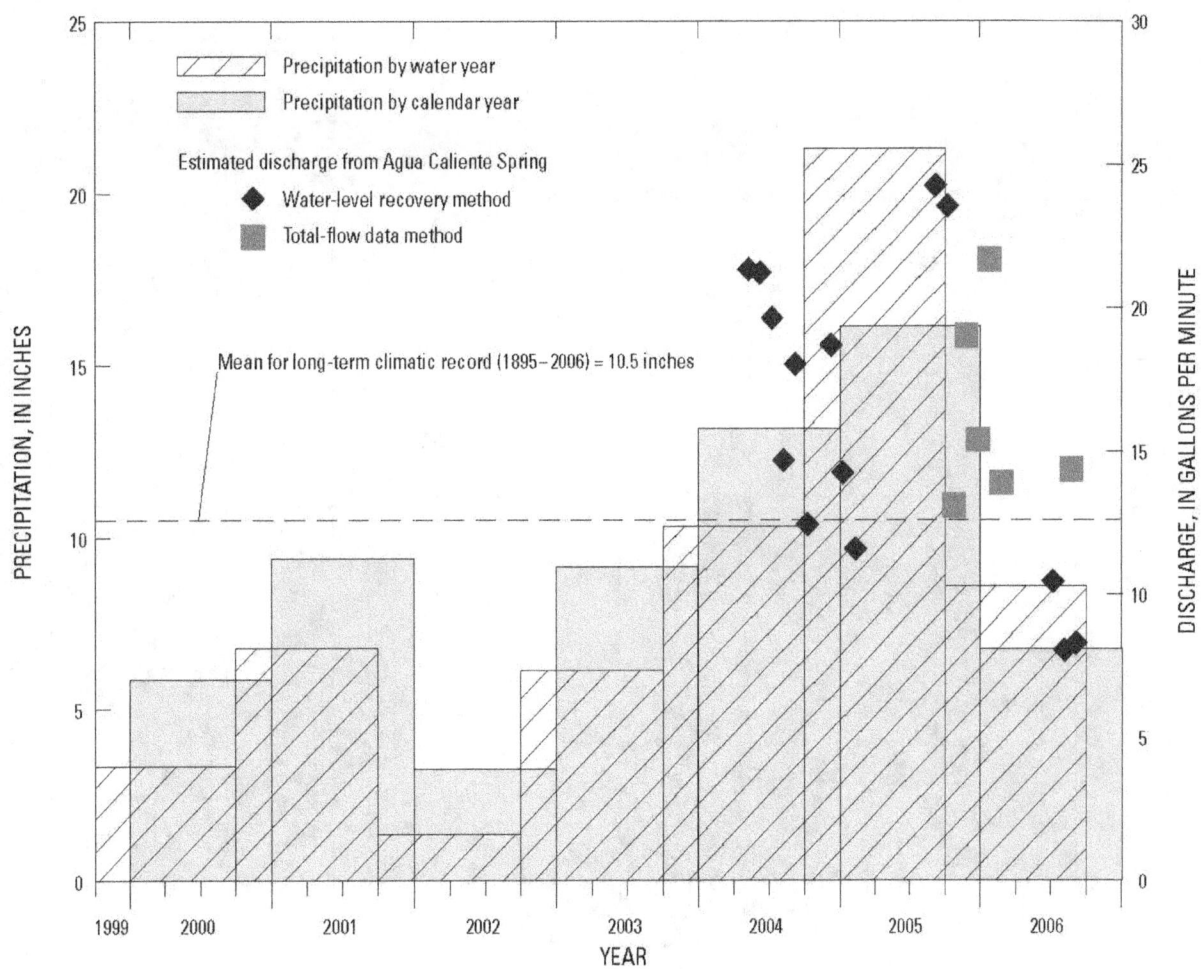

Figure 21.   Measured and estimated discharge, and simulated precipitation, for the (*A*) Agua Caliente, (*B*) Fenced, and (*C*) Chino Springs, California, 2000–06.

B

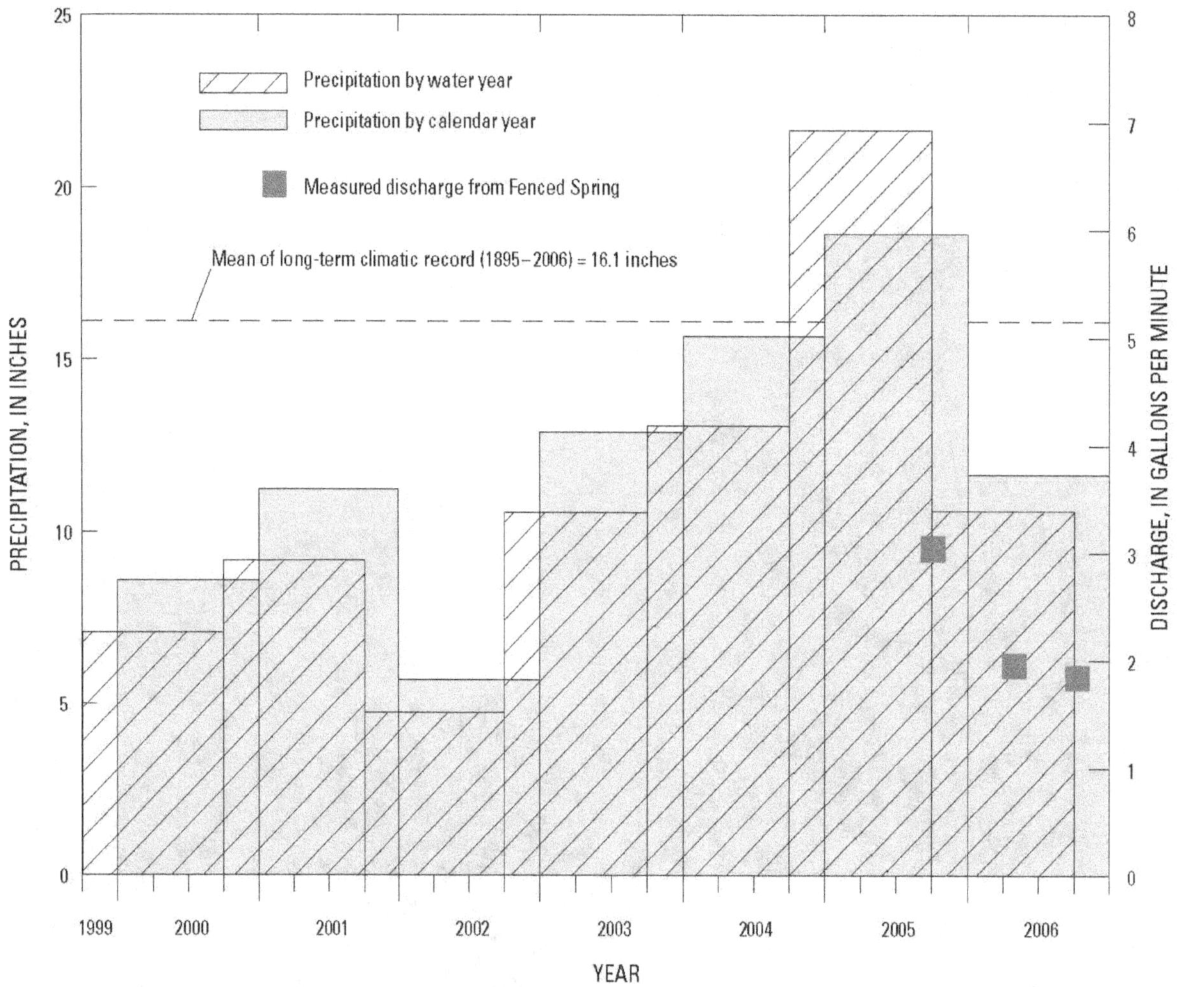

Figure 21.—Continued

C

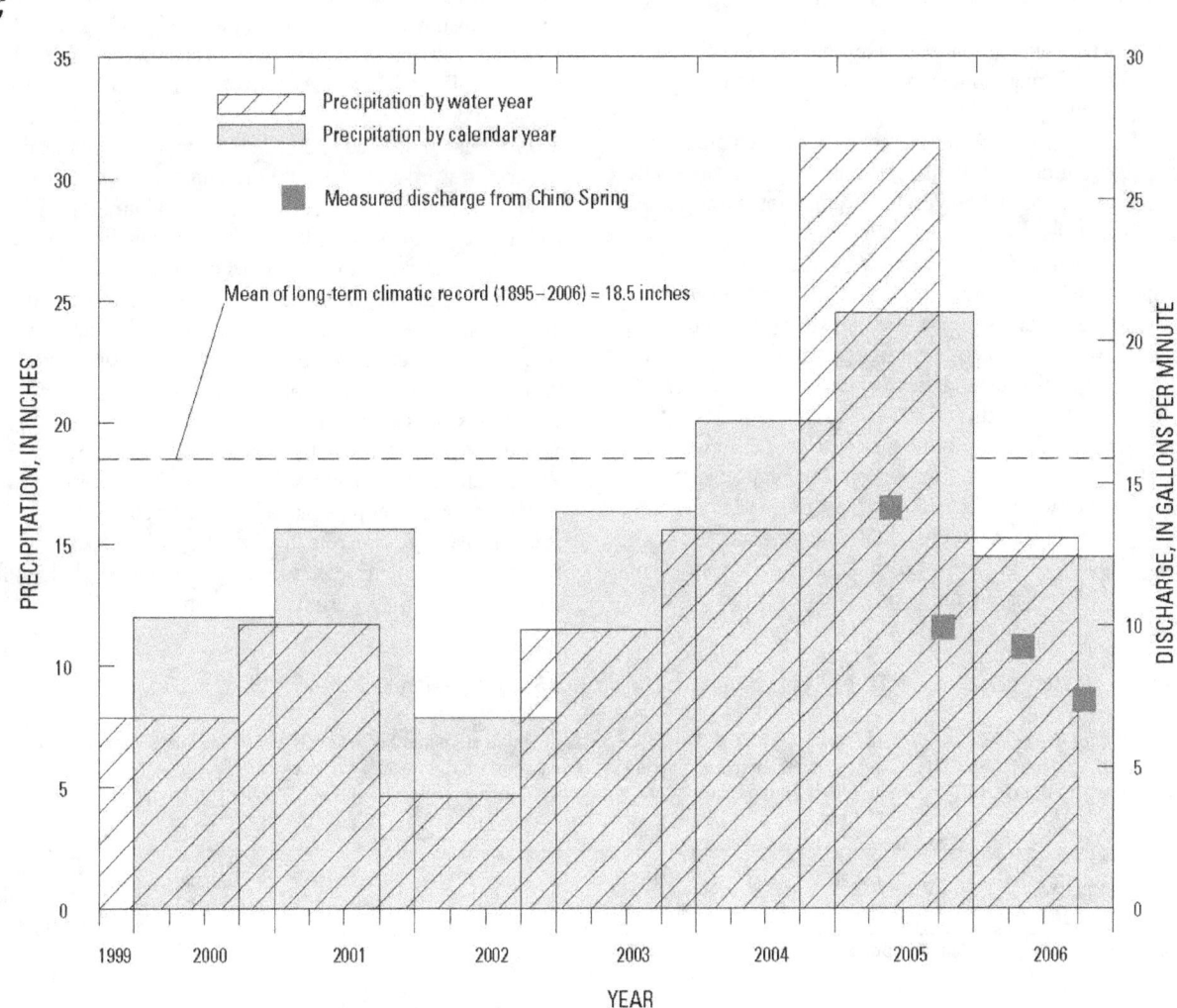

Figure 21.—Continued

Fenced, Chino Warm, and Chino Cold Springs

Instantaneous discharge was measured at Fenced Spring, Chino Warm Spring, and Chino Cold Spring during water-quality sampling events at the end of the wet and dry seasons (table 5). Discharge was measured by using the volumetric method, or a modification thereof. This method channels flow into a container of known volume over a known time (Rantz and others, 1982).

Use of a flume or weir plate would be the preferred way to measure discharge at Fenced Spring; however, tribal regulation restricts disturbance of the area around the spring. Thermal water at Fenced Spring seeps out into the bottom of a small pool that extends several feet across and is about 2 feet deep. Water in this pool discharges over its lowest bank, from which it continues to flow toward a stream at the bottom of Palm Canyon. In order to measure discharge, at least 35 gal of water was removed from the pool to depress its water level below where it would discharge. A known volume (2.5 or 5 gal) was then removed and time for the water level to recover to the same altitude on a measuring rod placed in the pool was recorded. The procedure was repeated several times to ensure a constant value (steady state) had been achieved. Instantaneous discharge measurements taken during three sampling trips, and associated measurement precision (standard deviation divided by mean expressed as percent), are shown in table 5. The table does not contain results from the first trip in April 2005 because discharge never stabilized— it declined continuously from 2.9 to 0.9 gal/min during 10 successive measurements. The reason that discharge at Fenced Spring failed to reach a steady state during this first of four visits to the site is unknown; however, a rather large area in the vicinity of the spring was wet following above-average precipitation, so shallow groundwater in soil adjacent to the pool may have seeped through the banks into the pool, and because this source was depleted during successive measurements, the discharge would have decreased.

Discharge from Chino Warm Spring flows vertically through a 2-in. steel pipe, makes a 90-degree bend, and pours into a large concrete tub. Discharge was obtained by measuring the time needed to fill a 2.5-gal bucket held at the end of the steel pipe. Results are shown in table 5.

Table 5.   Measured discharge at Fenced, Chino Warm, and Chino Cold Springs, California, 2005–06.

[See figure 1 for site locations. Abbreviations: USGS ID, U.S. Geological Survey identification number, which is the unique number for each site in the USGS NWIS (National Water Information System) database; mm/dd/yyyy, month/day/year; hh:mm, hour:minute; ft³/sec, cubic feet per second; gal/min, gallons per minute; RSD in percent, relative standard deviation in percent of mean discharge from multiple measurements (same as coefficient of variation)]

USGS ID	Local identifier	Date (mm/dd/yyyy)	Time (hh:mm)	Discharge, (ft³/sec)	Discharge, (gal/min)	Discharge precision, (RSD in percent)
334401116321201	Fenced Spring	09/08/2005	09:45	0.00677	3.04	2.6
		04/04/2006	10:30	0.00434	1.95	14
		09/12/2006	10:15	0.0041	1.84	3.2
335030116360401	Chino Warm Spring	04/28/2005	09:00	0.0314	14.1	3.3
		09/10/2005	10:00	0.022	9.87	0.8
		04/07/2006	14:00	0.0205	9.2	0.7
		09/13/2006	13:00	0.0163	7.32	1.4
335030116360601	Chino Cold Spring	04/07/2006	13:00	0.001084	0.486	10
		09/13/2006	10:30	0.00496	2.23	5

The nearby Chino Cold Spring is a very slow seep of groundwater at ambient temperatures in the Chino Canyon. A small amount of soil was excavated at the seep to permit the formation of a small pool, from which the flow was channeled into a small container to measure discharge. Water samples were collected from Chino Cold Spring only during the second year of this study, and discharge from two measurements in 2006 is given in table 5.

Discharge at both Fenced Spring and Chino Warm Spring was higher in 2005 than in 2006 (fig. 21B and C). Discharge from Chino Warm Spring was highest (14.1 gal/min) in the spring of 2005 and lowest (7.3 gal/min) at the end of summer in 2006. These few measurements, when compared to precipitation amounts calculated for their respective watershed areas (fig. 21B and C), suggest that spring discharge is positively correlated to recent precipitation intensity at Fenced Spring and Chino Warm Spring, as it also appears to be at Agua Caliente Spring. In order to establish a relation with greater degree of confidence, Fenced and Chino Warm Springs would need to be monitored over a longer period of several wet and dry cycles.

Spring Temperature

Water and air temperature were measured by using a hand-held conventional glass-bulb or electronic thermometer at each of the study sites during the planned water-quality sample collections. In addition, continuous temperature was measured at Agua Caliente Spring and Fenced Spring (fig. 22) using Hobo Water Temp Pro (Onset Inc., Pocasset, Massachusetts) probes. The temperature probes have an accuracy of ±0.2°C. At the Agua Caliente Spring, the temperature probe was attached to a cable and lowered into the pool to just above the spring's orifice. At Fenced Spring, the probe was placed at the bottom on one side of the Fenced Spring pool and fastened to the end of a 10-in. metal bracket (to permit ease of finding for subsequent retrieval), which was inserted deeply into the pool's bank, leaving the temperature sensor exposed at the water-bottom sediment interface.

Available records indicate that the temperature of the Agua Caliente Spring has varied over a narrow range during the past century, from a low of 37.8°C in 1917 to a high of 42.2°C in 1953 (table 4). Water temperature measurements collected at Agua Caliente Spring during this study were nearly constant, ranging only from a low of 40.7°C to a high of 41.8°C between April 29, 2005, and September 12, 2006, (fig. 22). Unlike the discharge of the Agua Caliente Spring, the temperature of the spring does not appear to be influenced by recent precipitation.

The apparent influence of precipitation on discharge from the Agua Caliente Spring and lack of influence of precipitation on the temperature of the spring can be explained by a piston-flow conceptual model for the spring, which is similar to the flow of water through a hose. Precipitation in the surrounding mountains recharges the underlying geothermal reservoir through fractures in the basement complex. The geothermal reservoir is large compared to its seasonal recharge, and is surrounded by low-permeability basement complex. Water slowly moves through the reservoir and is heated as it flows toward an exit, such as a continuous fracture or fault (fig. 10). The fractures or faults are relatively low-volume compared to the reservoir; therefore, as water recharges the reservoir, the discharge from the reservoir occurs almost simultaneously, similar to when water is added to a full hose. Because water movement through the reservoir is a long process, and the seasonal recharge pulses are small compared to the total water stored in the reservoir, the varying discharge from the reservoir maintains a relatively constant temperature.

The water temperature in Fenced Spring ranged from a low of 33.3°C to a high of 35.9°C between April 29, 2005, and July 20, 2006, (fig. 22). The temperature signal of Fenced Spring generally follows the sinusoidal seasonal cycle in regional air temperatures (fig. 22), although its range is markedly attenuated in the pool. The residence time of water in the small pool is never longer than a few hours, on the basis of measured discharges and the pool's dimensions, so it is not surprising that temperature in the pool reflects fluctuations in air temperature.

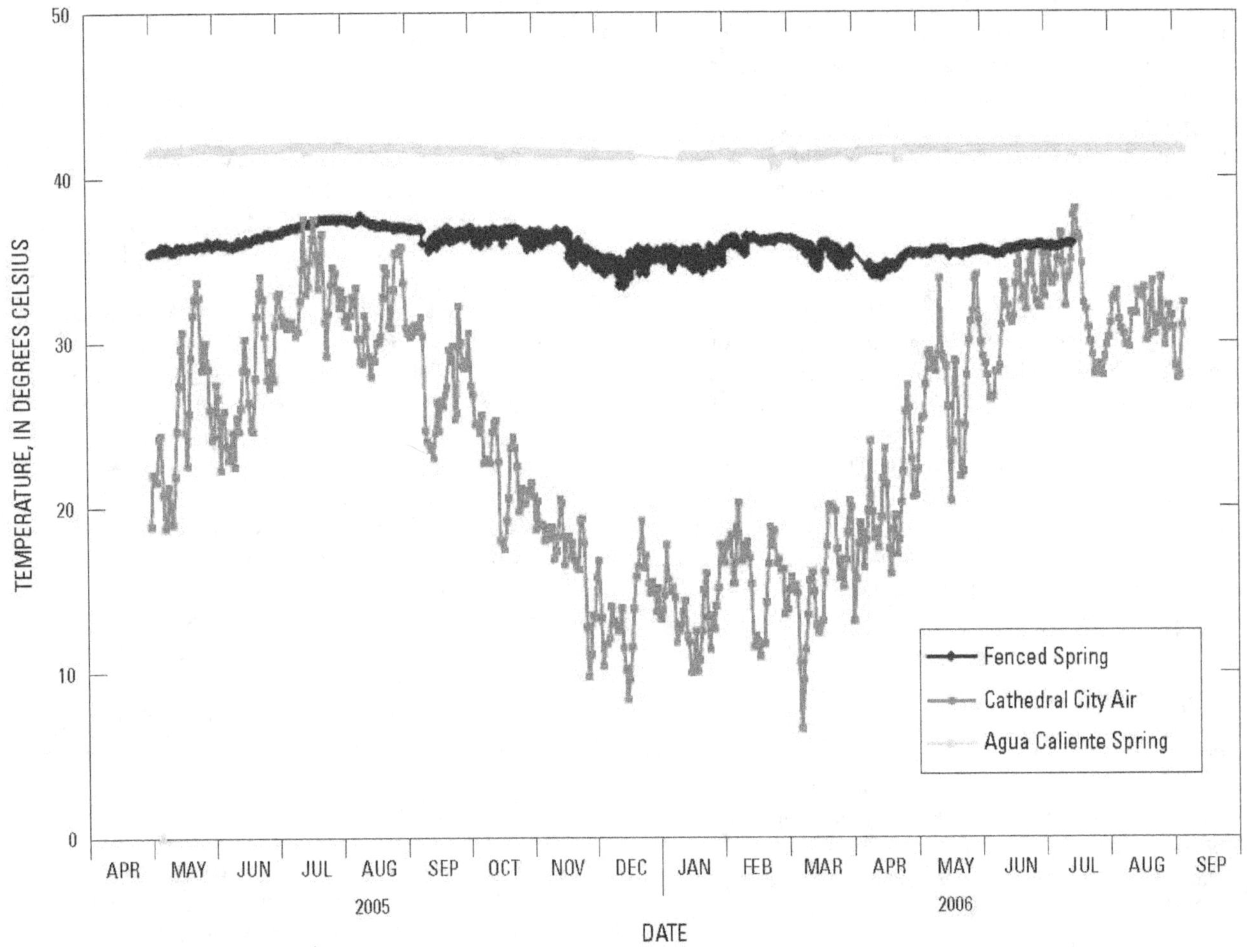

Figure 22.   Measured water temperature at the Agua Caliente and Fenced Springs, and air temperature at nearby Cathedral City, California, 2004–06.

Defining the Source and Age of Spring Discharge

By Roy A. Schroeder and Gregory A. Smith

Available historical water-quality data and seasonal data collected during this study were used to define the source(s) and the age(s) of water discharged by the Agua Caliente Spring and to ascertain the seasonal and longer-term variability of chemical characteristics of the spring discharge. Water-quality data collected from other springs and wells in the area were used to help evaluate the source of water to the Agua Caliente Spring.

Sample Collection, Laboratory Analysis, and Data Storage

Sample Collection

Water-quality samples were collected from Agua Caliente Spring, Fenced Spring, Chino Warm Spring, and Chino Cold Spring (fig. 1) during the fall and spring of 2005 and 2006. In addition, miscellaneous springs and wells were sampled on an irregular basis. Sampling procedures followed protocols described in the USGS field manual (U.S. Geological Survey, 2006), unless indicated otherwise, in which case additional details are provided below. Water samples were collected with a peristaltic pump and flexible C-flex tubing lowered to the bottom of pooled water formed by the spring's discharge. (Chino Warm Spring discharged through a galvanized pipe so the tubing was simply inserted into the pipe at this site.) A short stainless-steel tube was affixed to the inlet end of the flexible tubing as a weight to hold the intake in place, just above the bottom of the pool and near where the spring's orifice appeared to be located, while water was pumped to collection vessels. First, water for immediate determination of several constituents in the field was obtained. This was followed by collection of samples for analysis of constituents that required no filtration. Air and water temperature were determined in the field at the time sample collection began, and it is these initial temperatures that are archived with the sample's water-quality data, although air temperature increased markedly during the several hours that elapsed before completion of sample collection. Specific conductance, pH, alkalinity, dissolved oxygen, and dissolved hydrogen sulfide were measured in the field. Unfiltered samples for analysis of tritium (^3H) were collected in one-liter polyethylene bottles. Unfiltered samples for analysis of stable isotopes in water were collected in small glass vials. The stable isotope and tritium bottles were sealed with a conical plastic screw cap and taped with electrical tape to preclude leakage and evaporation prior to analysis.

Chromium speciation samples were obtained using a 10-milliliter (mL) syringe with an attached 0.45-micrometer (µm) disk filter. After the syringe was thoroughly rinsed and filled with sample water, 4 mL were forced through the disk filter, and the next 2 mL of the sample water were filtered slowly into a small centrifuge vial for analysis of total chromium. Hexavalent chromium, Cr (VI), then was collected by attaching a small cation exchange column to the syringe filter and, after conditioning the column with 2 mL of sample water, 2 mL were collected in a second centrifuge vial. Both vials were preserved with 10 microliters (µL) of 7.5 Normal (N) nitric acid (Ball and McCleskey, 2003).

After sufficient water for determination of field constituents and for distribution of aliquots into appropriate containers was obtained as described above, a 0.45-µm capsule filter (polyethersulfone membrane in a polypropylene housing) was attached to the outlet end of the flexible tubing, and filtered sample water was distributed into either polyethylene or amber glass bottles, depending on the constituent to be analyzed. Filtered water for analysis of anions and strontium isotopes was collected in polyethylene bottles. Filtered water for analysis of major and minor cations and trace elements was collected in polyethylene bottles and stabilized by addition of 6N nitric acid in the field to a pH of less than or equal to 2. The polyethylene bottle containing filtered water for determination of arsenic speciation was completely taped to exclude light and the sample was acidified by addition of 6N hydrochloric acid to a pH of less than or equal to 2. Dark brown polyethylene bottles were used for nutrients, and these samples were stored and shipped on ice to the laboratory, where they were refrigerated until analysis.

Water for analysis of "common" atmospheric gases was collected in a 150-mL glass bottle by inverting the bottle in a bucket of sample water, then allowing the peristaltic pump to deliver sufficient water to displace several volumes of water in the collection bottle. The bottle was sealed with a rubber septum, in which a syringe needle was inserted in order to allow a small amount of water to vent as the septum was pushed into the neck of the bottle while it was inverted and submerged in the bucket. Three replicates were taken, and they were stored and shipped on ice to the laboratory.

Filtered water for analysis of stable carbon isotopes and carbon-14 was collected by bottom filling an amber glass bottle that was first purged with three bottle volumes of sample water to ensure displacement of all atmospheric carbon dioxide prior to being sealed with a cap so that the sample had no head space.

Water for analysis of noble gases was sampled last and, for its collection, the C-flex tubing was replaced with Tygon tubing, which is less porous to gas exchange. An approximately 12-inch length of ⅜-inch inside diameter copper tubing was inserted in the Tygon between the inlet and the peristaltic pump head so that water was "pulled" through the copper tubing as the pump was running. The copper tube was held vertically in a manifold (a metal plate holder with two clamps) and tapped vigorously to dislodge any gas bubbles that might have adhered to the copper or to the Tygon. The pumping rate was then slowed as much as the pump would permit to minimize cavitation, or stripping of gases, and the outlet end of the copper tube was crimped shut as the water was being pumped until flow was completely stopped, after which the inlet end of the copper tube was also crimped shut, thereby trapping the sample water and its associated noble gases (Weiss, 1968). The copper tubing containing the water sample was then removed from the manifold and placed in bubble pack and urethane padding to minimize vibration during shipment to the laboratory.

Laboratory Analysis

Several laboratories performed chemical and isotopic analyses on water samples collected for this study. Major ions, minor and trace elements, and nutrients were analyzed by the USGS National Water-Quality Laboratory (NWQL; Timme, 1995) by various methods as described in Faires (1993), Fishman and Friedman (1989), Fishman (1993), Garbarino (1999), Garbarino and others (2006), McLain (1993), Patton and Krysalla (2003), Patton and Truitt (1992, 2000), and Struzeski and others (1996). Chromium and arsenic speciation was determined by the USGS National Research Program (NRP) laboratory in Boulder, Colorado, using methods described in Ball and McClesky (2003) and McClesky and others (2003). Tritium activity was measured by the NRP Tritium-Light Isotope laboratory in Menlo Park, California, by electrolytic enrichment using glass cells with electodes of nickel and stainless steel (Ostlund and Werner, 1962). The electrolyzed samples were then counted in liquid scintillation counters as 1:1 mixtures of water and a commercial scintillator (Thatcher and others, 1977). Helium-3 and other noble gases were analyzed by Lawrence Livermore National Laboratory (LLNL) by accelerator-mass spectrometry (Eaton and others, 2004). Dissolved "common" atmospheric gases were analyzed by the NRP chlorofluorocarbon (CFC) laboratory in Reston, Virginia, using gas chromatography (N. Plummer, USGS, written commun.). Free gas (bubbles) from Agua Caliente Spring was captured by an inverted glass funnel lowered into the spring's pool that was attached with tubing to an evacuated glass bulb at the surface and analyzed by gas chromatography (R. Mariner, USGS, written commun.). Stable hydrogen and oxygen isotopes of water were analyzed by the NRP stable-isotope laboratory in Reston, Virginia (Epstein and Mayeda, 1953; Coplen and others, 1991; Coplen, 1994). Strontium isotopes were analyzed by the NRP heavy-isotope laboratory in Menlo Park, California, by using methods described in Bullen and others (1996). Stable carbon isotopes and carbon-14 were analyzed at the University of Waterloo (Donahue and others, 1990; Jull and others, 2004).

Data Storage

All analytical results obtained for this study from the NWQL and measurements made in the field are stored in NWIS (National Water Information System), which is the USGS computer database for storage of water information. Concentration of each analyte or water property is stored under a unique 5-digit parameter code, and each site is designated by a unique 8- or 15-digit station identification number (Site ID). Selected data obtained from labs other than the NWQL also are stored in NWIS. The information in NWIS can be accessed by the public at http://waterdata.usgs.gov/nwis/.

Water-Quality Results

Available historical water-quality data and seasonal data collected during this study were used to describe the general chemistry of water discharged from Agua Caliente Spring and other selected springs and wells in the study area. These data were used to define saturation controls on major-ion composition, temporal trends, the source of anions, and the temperature of the geothermal source water. The stable isotopes of oxygen and hydrogen were used to estimate the altitude of recharge for the groundwater samples, and the dissolved-gas concentration was used to estimate the temperature of recharge. The radioactive isotopes of carbon and hydrogen were used to estimate the age of the water samples.

General Chemistry

The general chemistry discussed in this report includes field data (pH, specific conductance, dissolved oxygen, and hydrogen sulfide), major ions, nutrients, and trace elements.

Field Data

Specific conductance, pH, dissolved oxygen, and hydrogen sulfide were measured in the field prior to the collection of all samples collected for this study (tables 6 and 7). The pH of most natural waters is slightly alkaline (pH slightly greater than 7) as a result of equilibration with the calcium carbonate-bicarbonate dissolution/precipitation process. Substantially higher pH (greater than 9) indicates the bicarbonate buffering capacity is being exceeded. Water high in pH also has a relatively high sodium-calcium ratio and low calcium concentration because the solubility of calcium carbonate decreases with increasing pH. Current and historical samples from Agua Caliente Spring exhibit these characteristics, indicating groundwater that has undergone substantial change during subsurface flow that distinguishes it from nonthermal regional groundwater (fig. 23A; tables 4 and 6).

Table 6. Analyses of water samples collected for the Agua Caliente Spring study, California, study, 2005–07.

[See figures 1 and 2 for site locations. The five-digit USGS parameter code in parentheses is used to uniquely identify each water constituent or property in the NWIS database. **Abbreviations:** USGS ID, U.S. Geological Survey identification number, which is the unique number for each site in the USGS NWIS (National Water Information System) database; mm/dd/yyyy, month/day/year; hh:mm, hour: minute; °C, degrees Celsius; µS/cm, microsiemens per centimeter at 25°C; mg/L, milligrams per liter; µg/L, micrograms per liter; E, value estimated; CaCO₃, calcium carbonate; TU, tritium units; <, less than; –, no data]

Local identifier	USGS ID	Date (mm/dd/yyyy)	Time (hh:mm)	Calcium, water, filtered (mg/L) (00915)	Magnesium, water, filtered (mg/L) (00925)	Potassium, water, filtered (mg/L) (00935)	Sodium, water, filtered (mg/L) (00930)	Bromide, water, filtered (mg/L) (71870)	Chloride, water, filtered (mg/L) (00940)	Fluoride, water, filtered (mg/L) (00950)
Agua Caliente Spring	334924116324301	01/28/2005	11:30	–	–	–	–	–	–	–
		04/27/2005	08:00	1.3	<0.008	0.6	67.5	0.08	22.5	2.68
		09/09/2005	09:00	1.35	<0.008	0.59	65	0.08	23.4	2.78
		04/05/2006	10:00	1.3	E0.004	0.62	66.6	0.09	22.8	2.77
		09/11/2006	10:00	1.23	E0.004	0.63	65	0.08	22.3	2.64
Hot Spring	334355116320601	01/28/2005	13:30	–	–	–	–	–	–	–
Fenced Spring	334401116321201	01/28/2005	13:15	–	–	–	–	–	–	–
		04/29/2005	10:00	3.02	0.022	1.09	104	0.23	45.2	2.62
		09/08/2005	09:45	3.11	0.011	1	101	0.22	46	2.87
		04/04/2006	10:30	2.9	0.013	0.88	100	0.23	45.7	2.81
		09/12/2006	10:15	2.8	0.018	0.84	99.2	0.22	45.1	2.79
Trading Post Spring	334413116322001	01/28/2005	14:00	–	–	–	–	–	–	–
Chino Warm Spring	335030116360401	04/28/2005	09:00	1.22	E0.004	0.77	45.3	E0.02	4.23	1.07
		09/10/2005	10:00	1.18	E0.006	0.85	43.2	E0.01	4.16	1.09
		04/07/2006	14:00	–	–	–	–	–	–	–
		09/13/2006	13:00	–	–	–	–	–	–	–
Chino Cold Spring	335030116360601	04/28/2005	12:00	–	–	–	–	–	–	–
		09/10/2005	13:00	73.2	6.14	6.49	16.3	0.05	7.09	0.2
		04/07/2006	13:00	73.8	6.64	7.79	17.5	0.05	7.32	0.25
		09/13/2006	10:30	70.3	6.1	6.59	16.9	0.05	7.03	0.2
Dos Palmas Spring	334314116333101	09/22/2005	09:30	–	–	–	–	–	–	–
Indian Spring	334317116331801	09/22/2005	11:00	–	–	–	–	–	–	–
Cedar Spring	334037116343801	01/26/2006	09:55	–	–	–	–	–	–	–
004S004E14E002S (OW-1 #2)	334924116324402	08/01/2007	09:15	1.21	0.13	0.68	70.7	–	23.5	2.75
004S004E14E003S (OW-1 #3)	334924116324403	08/01/2007	10:45	1.05	0.496	0.89	70.7	–	22.8	2.76
004S004E14E004S (OW-2)	334923116324401	08/01/2007	12:50	1.62	<0.014	0.74	68.5	–	22.7	2.68
004S004E14Q001S	334905116320901	08/24/2006	10:00	28.1	1.44	3.56	32.5	0.05	16.1	0.28
Chino Canyon Creek	10257720	04/07/2006	15:00	–	–	–	–	–	–	–

Table 6. Analyses of water samples collected for the Agua Caliente Spring, California, study, 2005–07.—Continued

[See figures 1 and 2 for site locations. The five-digit USGS parameter code in parentheses is used to uniquely identify each water constituent or property in the NWIS database. **Abbreviations:** USGS ID; U.S. Geological Survey identification number, which is the unique number for each site in the USGS NWIS (National Water Information System) database; mm/dd/yyyy, month/day/year; hh:mm, hour: minute; °C, degrees Celsius; µS/cm, microsiemens per centimeter at 25°C; mg/L, milligrams per liter; µg/L, micrograms per liter; E, value estimated; CaCO₃, calcium carbonate; TU, tritium units; <, less than; –, no data]

Local identifier	USGS ID	Date (mm/dd/yyyy)	Time (hh:mm)	Silica, water, filtered (mg/L) (00955)	Sulfate, water, filtered (mg/L) (00945)	Residue on evaporation, dried at 180 °C, water, filtered (mg/L) (70300)	Ammonia plus organic nitrogen, water, filtered (mg/L as nitrogen) (00623)	Ammonia, water, filtered (mg/L as nitrogen) (00608)	Nitrite plus nitrate, water, filtered (mg/L as nitrogen) (00631)	Nitrite, water, filtered (mg/L as nitrogen) (00613)	Orthophosphate, water, filtered (mg/L as phosphorus) (00671)
Agua Caliente Spring	334924116324301	01/28/2005	11:30	–	–	–	–	–	–	–	–
		04/27/2005	08:00	48.4	25.3	234	E0.07	<0.04	<0.06	<0.008	<0.02
		09/09/2005	09:00	50.9	28	237	0.12	<0.04	<0.06	<0.008	<0.02
		04/05/2006	10:00	53.5	28.2	228	E0.07	<0.04	<0.06	<0.008	<0.02
		09/11/2006	10:00	48.8	26.7	221	<0.1	0.038	<0.06	<0.002	0.011
Hot Spring	334355116320601	01/28/2005	13:30	–	–	–	–	–	–	–	–
Fenced Spring	334401116321201	01/28/2005	13:15	–	–	–	–	–	–	–	–
		04/29/2005	10:00	30.5	97.8	335	E0.07	<0.04	<0.06	<0.008	<0.02
		09/08/2005	09:45	32.2	100	336	E0.09	<0.04	<0.06	<0.008	<0.02
		04/04/2006	10:30	33.3	99.9	325	<0.1	<0.04	<0.06	<0.008	<0.02
		09/12/2006	10:15	30.8	98	324	E0.08	E0.007	<0.06	<0.002	0.008
Trading Post Spring	334413116322001	01/28/2005	14:00	–	–	–	–	–	–	–	–
Chino Warm Spring	335030116360401	04/28/2005	09:00	55.3	10.2	186	<0.1	<0.04	<0.06	<0.008	<0.02
		09/10/2005	10:00	58.4	9.57	177	–	–	–	–	–
		04/07/2006	14:00	–	–	–	–	–	–	–	–
		09/13/2006	13:00	–	–	–	–	–	–	–	–
Chino Cold Spring	335030116360601	04/28/2005	12:00	–	–	–	–	–	–	–	–
		09/10/2005	13:00	28.1	52.1	309	E0.08	<0.04	1.53	<0.008	<0.02
		04/07/2006	13:00	28.6	56.9	327	0.25	<0.04	1.99	E0.004	<0.02
		09/13/2006	10:30	26.2	52.1	300	0.13	<0.01	2.16	0.003	0.012
Dos Palmas Spring	334314116333101	09/22/2005	09:30	–	–	–	–	–	–	–	–
Indian Spring	334317116331801	09/22/2005	11:00	–	–	–	–	–	–	–	–
Cedar Spring	334037116343801	01/26/2006	09:55	–	–	–	–	–	–	–	–
004S004E14E002S (OW-1 #2)	334924116324402	08/01/2007	09:15	51.9	35.2	–	–	0.048	<0.06	<0.002	–
004S004E14E003S (OW-1 #3)	334924116324403	08/01/2007	10:45	48.6	26.5	–	–	0.161	<0.06	E0.003	–
004S004E14E004S (OW-2)	334923116324401	08/01/2007	12:50	63.5	26.1	–	–	0.051	<0.06	<0.002	–
004S004E14Q001S	334905116320901	08/24/2006	10:00	20.4	25.3	214	E0.08	E0.005	1.72	<0.002	0.018
Chino Canyon Creek	10257720	04/07/2006	15:00	–	–	–	–	–	–	–	–

Table 6. Analyses of water samples collected for the Agua Caliente Spring, California, study, 2005–07.—Continued

[See figures 1 and 2 for site locations. The five-digit USGS parameter code in parentheses is used to uniquely identify each water constituent or property in the NWIS database. **Abbreviations:** USGS ID, U.S. Geological Survey identification number, which is the unique number for each site in the USGS NWIS (National Water Information System) database; mm/dd/yyyy, month/day/year; hh:mm, hour: minute; °C, degrees Celsius; µS/cm, microsiemens per centimeter at 25°C; mg/L, milligrams per liter; µg/L, micrograms per liter; E, value estimated; $CaCO_3$, calcium carbonate; TU, tritium units; <, less than; –, no data]

Local identifier	USGS ID	Date (mm/dd/yyyy)	Time (hh:mm)	Trace elements									
				Arsenic, water, filtered (µg/L) (01000)	Barium, water, filtered (µg/L) (01005)	Boron, water, filtered (µg/L) (01020)	Iron, water, filtered (µg/L) (01046)	Lithium, water, filtered (µg/L) (01130)	Manganese, water, filtered (µg/L) (01056)	Selenium, water, filtered (µg/L) (01145)	Strontium, water, filtered (µg/L) (01080)	Vanadium, water, filtered (µg/L) (01085)	Uranium (natural), water, filtered (µg/L) (22703)
Agua Caliente Spring	3349241116324301	01/28/2005	11:30	–	–	–	–	–	–	–	–	–	–
		04/27/2005	08:00	E0.2	<1	144	E4	32	E0.5	1.8	7	<2	<0.04
		09/09/2005	09:00	<0.12	<1	143	<6	29	<0.6	<0.08	7.4	<0.1	<0.04
		04/05/2006	10:00	<0.12	E0.5	136	E4	34	0.7	E0.07	7.5	<0.1	<0.04
		09/11/2006	10:00	<0.12	<1	146	<6	31	E0.4	<0.08	7.8	<0.1	<0.04
Hot Spring	3343551116320601	01/28/2005	13:30	–	–	–	–	–	–	–	–	–	–
Fenced Spring	3344011116321201	01/28/2005	13:15	–	–	–	–	–	–	–	–	–	–
		04/29/2005	10:00	2.5	E0.8	169	<6	75	E0.6	0.6	7.3	<2	<0.04
		09/08/2005	09:45	2.3	<1	166	<6	68	E0.4	<0.08	7.1	E0.06	<0.04
		04/04/2006	10:30	2.1	E0.7	159	<6	83	E0.5	<0.08	6.9	E0.06	E0.04
		09/12/2006	10:15	2.2	E0.7	169	<6	78	E0.5	<0.08	7.3	<0.1	<0.04
Trading Post Spring	3344131116322001	01/28/2005	14:00	–	–	–	–	–	–	–	–	–	–
Chino Warm Spring	3350301116360401	04/28/2005	09:00	0.7	<1	68	<6	15	<0.6	E0.3	9.1	<2	E0.02
		09/10/2005	10:00	<0.12	2.9	67	<6	14	<0.6	<0.08	8.7	<0.1	E0.02
		04/07/2006	14:00	–	–	–	–	–	–	–	–	–	–
		09/13/2006	13:00	–	–	–	–	–	–	–	–	–	–
Chino Cold Spring	3350301116360601	04/28/2005	12:00	–	–	–	–	–	–	–	–	–	–
		09/10/2005	13:00	E0.11	54.9	32	<6	11	<0.6	1.3	296	4.9	21.1
		04/07/2006	13:00	0.12	61.1	31	8	12	4.7	1.1	334	4.7	22.6
		09/13/2006	10:30	0.16	53.7	31	<6	11	0.9	1.1	313	5.6	22.1
Dos Palmas Spring	3343141116333101	09/22/2005	09:30	–	–	–	–	–	–	–	–	–	–
Indian Spring	3343171116331801	09/22/2005	11:00	–	–	–	–	–	–	–	–	–	–
Cedar Spring	3343071116343801	01/26/2006	09:55	–	–	–	–	–	–	–	–	–	–
004S004E14E002S (OW-1 #2)	3349241116324402	08/01/2007	09:15	–	–	149	8	–	0.4	–	–	–	–
004S004E14E003S (OW-1 #3)	3349241116324403	08/01/2007	10:45	–	–	163	1,180	–	18.5	–	–	–	–
004S004E14E004S (OW-2)	3349231116324401	08/01/2007	12:50	–	–	156	10	–	0.4	–	–	–	–
004S004E14Q001S	3349051116320901	08/24/2006	10:00	0.56	72	27	<6	4	<0.6	0.65	245	20	4.16
Chino Canyon Creek	10257720	04/07/2006	15:00	–	–	–	–	–	–	–	–	–	–

Table 6. Analyses of water samples collected for the Agua Caliente Spring, California, study, 2005–07.—Continued

[See figures 1 and 2 for site locations. The five-digit USGS parameter code in parentheses is used to uniquely identify each water constituent or property in the NWIS database. Abbreviations: USGS ID, U.S. Geological Survey identification number, which is the unique number for each site in the USGS NWIS (National Water Information System) database; mm/dd/yyyy, month/day/year; hh:mm, hour: minute; °C, degrees Celsius; µS/cm, microsiemens per centimeter at 25°C; mg/L, milligrams per liter; µg/L, micrograms per liter; E, value estimated; CaCO$_3$, calcium carbonate; TU, tritium units; <, less than; –, no data]

Local identifier	USGS ID	Date (mm/dd/yyyy)	Time (hh:mm)	Dissolved oxygen, water, unfiltered, (mg/L) (00300)	pH, water, unfiltered, field (standard units) (00400)	Specific conductance, water (µS/cm at 25°C) (00095)	Temperature, water (°C) (00010)	Alkalinity, water, filtered, fixed endpoint (pH 4.5) titration, field (mg/L as CaCO$_3$) (39036)	Alkalinity, water, filtered, incremental titration, field (mg/L as CaCO$_3$) (39086)	Bicarbonate, water, filtered, incremental titration, field (mg/L) (00453)	Carbonate, water, filtered, incremental titration, field (mg/L) (00452)	Hydrogen sulfide, water, unfiltered (mg/L) (71875)
Agua Caliente Spring	334924116324301	01/28/2005	11:30	–	–	330	–	–	–	–	–	1.6
		04/27/2005	08:00	<1	9.8	323	41.2	–	85	63	19	1.9
		09/09/2005	09:00	<0.1	9.7	342	41	–	88	63	20	2.5
		04/05/2006	10:00	2.1	9.7	336	41	–	79	51	20	2.1
		09/11/2006	10:00	0.6	9.7	328	41.7	86	85	59	19	–
Hot Spring	334355116320601	01/28/2005	13:30	–	–	522	–	–	–	–	–	–
Fenced Spring	334401116321201	01/28/2005	13:15	–	–	518	–	–	–	–	–	–
		04/29/2005	10:00	<1	8.8	504	31.2	–	58	65	3	0.1
		09/08/2005	09:45	2.1	9	532	33	–	61	68	3	0.3
		04/04/2006	10:30	1.3	8.8	528	32	–	62	68	4	0.4
		09/12/2006	10:15	1.3	8.8	505	32.5	60	61	67	3	0.3
Trading Post Spring	334413116322001	01/28/2005	14:00	–	–	508	–	–	–	–	–	–
Chino Warm Spring	335030116360401	04/28/2005	09:00	<1	9.8	213	40.8	–	77	50	19	0.1
		09/10/2005	10:00	2.4	9.8	219	41.5	–	84	59	20	0.2
		04/07/2006	14:00	0.6	9.7	218	40.7	–	–	–	–	0.1
		09/13/2006	13:00	1.6	9.9	264	41.3	–	–	–	–	0.1
Chino Cold Spring	335030116360601	04/28/2005	12:00	–	7.7	479	21.6	–	175	213	0	–
		09/10/2005	13:00	–	7.7	471	22	–	182	221	0	–
		04/07/2006	13:00	3.2	7.3	484	20.7	–	170	207	0	<0.01
		09/13/2006	10:30	4.8	7.4	475	22.4	160	165	201	0	<0.01
Dos Palmas Spring	334314116333101	09/22/2005	09:30	–	–	–	–	–	–	–	–	–
Indian Spring	334317116331801	09/22/2005	11:00	–	–	–	–	–	–	–	–	–
Cedar Spring	334037116343801	01/26/2006	09:55	–	–	–	–	–	–	–	–	–
004S004E14E002S (OW-1 #2)	334924116324402	08/01/2007	09:15	0.8	9.7	325	40.9	–	–	–	–	–
004S004E14E003S (OW-1 #3)	334924116324403	08/01/2007	10:45	0.9	9.6	327	39.8	–	–	–	–	–
004S004E14E004S (OW-2)	334923116324401	08/01/2007	12:50	2.1	9.6	328	40.7	–	–	–	–	–
004S004E14Q001S	334905116320901	08/24/2006	10:00	6.3	8	316	26.3	99	98	117	0	–
Chino Canyon Creek	102577720	04/07/2006	15:00	–	–	–	–	–	–	–	–	–

Table 6. Analyses of water samples collected for the Agua Caliente Spring, California, study, 2005–07.—Continued

[See figures 1 and 2 for site locations. The five-digit USGS parameter code in parentheses is used to uniquely identify each water constituent or property in the NWIS database. **Abbreviations:** USGS ID, U.S. Geological Survey identification number, which is the unique number for each site in the USGS NWIS (National Water Information System) database; mm/dd/yyyy, month/day/year; hh:mm, hour:minute; °C, degrees Celsius; µS/cm, microsiemens per centimeter at 25°C; mg/L, milligrams per liter; µg/L, micrograms per liter; E, value estimated; CaCO$_3$, calcium carbonate; TU, tritium units; <, less than; –, no data]

Local identifier	USGS ID	Date (mm/dd/yyyy)	Time (hh:mm)	Carbon-14, water, filtered (percent modern) (49933)	Carbon-13/ Carbon-12 ratio, water, unfiltered, (per mil) (82081)	Carbon-14 age uncorrected (years)	Deuterium/ Protium ratio, water unfiltered, (per mil) (82082)	Oxygen-18/ Oxygen-16 ratio, water unfiltered, (per mil) (82085)	Tritium (in TU)	Tritium, precision 1 sigma	Strontium 87/86
							Radioactive and stable isotopes				
Agua Caliente Spring	334924116324301	01/28/2005	11:30	–	–	–	–77.9	–10.96	–0.04	0.08	0.71055
		04/27/2005	08:00	–	–	–	–78.6	–10.9	0.03	0.09	0.71058
		09/09/2005	09:00	16.01	–11.75	15,149	–78.2	–10.84	0.04	0.11	–
		04/05/2006	10:00	17.95	–5.77	14,204	–77.9	–10.81	0.12	0.07	–
		09/11/2006	10:00	17.29	–10.01	14,513	–	–	0.04	0.09	–
Hot Spring	334355116320601	01/28/2005	13:30	–	–	–	–	–	–	–	–
Fenced Spring	334401116321201	01/28/2005	13:15	–	–	–	–72.9	–9.93	0	0.08	0.71215
		04/29/2005	10:00	36.85	–15.73	8,256	–71.7	–9.91	–0.15	0.13	0.71220
		09/08/2005	09:45	38.43	–16.27	7,908	–72.3	–9.91	–0.13	0.08	–
		04/04/2006	10:30	42.79	–15.98	7,020	–72.4	–9.94	–0.05	0.07	–
		09/12/2006	10:15	–	–	–	–	–	0.09	0.09	–
Trading Post Spring	334413116322001	01/28/2005	14:00	–	–	–	–	–	–0.04	0.08	–
Chino Warm Spring	335030116360401	04/28/2005	09:00	–	–	–	–81.1	–11.66	–0.07	0.13	0.71120
		09/10/2005	10:00	29.84	–11.27	10,000	–80.4	–11.63	0	0.12	0.71116
		04/07/2006	14:00	–	–	–	–	–	0.01	0.08	–
		09/13/2006	13:00	–	–	–	–	–	0.12	0.1	–
Chino Cold Spring	335030116360601	04/28/2005	12:00	–	–	–	–78.6	–11.01	1.58	0.14	–
		09/10/2005	13:00	93.95	–13.8	516	–78.1	–10.91	1.61	0.11	–
		04/07/2006	13:00	98.53	–15.28	122	–77.8	–11.02	1.52	0.1	–
		09/13/2006	10:30	101.5	–14.28	–123	–77.5	–10.98	0.94	0.11	0.71066
Dos Palmas Spring	334314116333101	09/22/2005	09:30	–	–	–	–68.7	–9.3	–	–	–
Indian Spring	334317116331801	09/22/2005	11:00	–	–	–	–68.1	–9.07	–	–	–
Cedar Spring	334037116343801	01/26/2006	09:55	–	–	–	–66.3	–10.06	–	–	–
004S004E14E002S (OW-1 #2)	334924116324402	08/01/2007	09:15	–	–	–	–78.1	–10.91	–	–	–
004S004E14E003S (OW-1 #3)	334924116324403	08/01/2007	10:45	–	–	–	–77.4	–10.91	–	–	–
004S004E14E004S (OW-2)	334923116324401	08/01/2007	12:50	–	–	–	–77.3	–10.57	–	–	–
004S004E14Q001S	334905116320901	08/24/2006	10:00	91.41	–12.5	743	–75.9	–10.76	1.75	0.13	0.71077
Chino Canyon Creek	10257720	04/07/2006	15:00	–	–	–	–80.7	–11.55	–	–	–

Table 6. Analyses of water samples collected for the Agua Caliente Spring, California, study, 2005–07.—Continued

[See figures 1 and 2 for site locations. The five-digit USGS parameter code in parentheses is used to uniquely identify each water constituent or property in the NWIS database. **Abbreviations:** USGS ID, U.S. Geological Survey identification number, which is the unique number for each site in the USGS NWIS (National Water Information System) database; mm/dd/yyyy, month/day/year; hh:mm, hour: minute; °C, degrees Celsius; µS/cm, microsiemens per centimeter at 25°C; mg/L, milligrams per liter; µg/L, micrograms per liter; E, value estimated; CaCO₃, calcium carbonate; TU, tritium units; <, less than; –, no data]

Local identifier	USGS ID	Date (mm/dd/yyyy)	Time (hh:mm)	Arsenic, total, (µg/L)	Arsenic (III) (µg/L)	Arsenic (V) [1] (µg/L)	Chromium, total, (µg/L)	Chromium (VI) (µg/L)	Chromium (III) [1] (µg/L)
Agua Caliente Spring	334924116324301	01/28/2005	11:30	–	–	–	–	–	–
		04/27/2005	08:00	0.5	<2	<0.5	<1	<1	<1
		09/09/2005	09:00	–	–	–	–	–	–
		04/05/2006	10:00	0.6	<1	<0.6	<1	<1	<1
		09/11/2006	10:00	–	–	–	–	–	–
Hot Spring	334355116320601	01/28/2005	13:30	–	–	–	–	–	–
Fenced Spring	334401116321201	01/28/2005	13:15	–	–	–	–	–	–
		04/29/2005	10:00	2.2	<2	>0.2	<1	<1	<1
		09/08/2005	09:45	–	–	–	–	–	–
		04/04/2006	10:30	2	1.2	<1	<1	<1	<1
		09/12/2006	10:15	–	–	–	–	–	–
Trading Post Spring	334413116322001	01/28/2005	14:00	–	–	–	–	–	–
Chino Warm Spring	335030116360401	04/28/2005	09:00	<0.5	<2	<0.5	<1	<1	<1
		09/10/2005	10:00	–	–	–	–	–	–
		04/07/2006	14:00	–	–	–	–	–	–
		09/13/2006	13:00	–	–	–	–	–	–
Chino Cold Spring	335030116360601	04/28/2005	12:00	–	–	–	–	–	–
		09/10/2005	13:00	–	–	–	–	–	–
		04/07/2006	13:00	<0.5	<1	<0.5	<1	<1	<1
		09/13/2006	10:30	–	–	–	–	–	–
Dos Palmas Spring	334314116333101	09/22/2005	09:30	–	–	–	–	–	–
Indian Spring	334317116331801	09/22/2005	11:00	–	–	–	–	–	–
Cedar Spring	334037116343801	01/26/2006	09:55	–	–	–	–	–	–
004S004E14E002S (OW-1 #2)	334924116324402	08/01/2007	09:15	–	–	–	–	–	–
004S004E14E003S (OW-1 #3)	334924116324403	08/01/2007	10:45	–	–	–	–	–	–
004S004E14E004S (OW-2)	334923116324401	08/01/2007	12:50	–	–	–	–	–	–
004S004E14Q001S	334905116320901	08/24/2006	10:00	–	–	–	–	–	–
Chino Canyon Creek	10257720	04/07/2006	15:00	–	–	–	–	–	–

[1] Calculated by difference.

Table 7. Historical and recent water-quality data from three production wells near the Agua Caliente Spring, California.

[See figure 1 for site locations. State well no., see well-numbering diagram in text; the five-digit USGS parameter code in parentheses is used to uniquely identify each water constituent or property in the NWIS database. **Abbreviations:** USGS ID, U.S. Geological Survey identification number which is the unique number for each site in the USGS NWIS (National Water Information System) database; mm/dd/ yyyy, month/day/year; °C, degrees Celsius; µS/cm, microsiemens per centimeter at 25°C; mg/L, milligrams per liter; µg/L, micrograms per liter; ft, feet; ft BLS, feet below land surface datum; <, less than; M, presence verified but not quantified; –, no data]

State well no.	USGS ID	Date (mm/dd/yyyy)	Temperature, water (°C) (00010)	Specific conductance, water, unfiltered (µS/cm) (00095)	pH, water, unfiltered, field (standard units) (00400)	Acid neutralizing capacity, water, unfiltered, fixed endpoint (pH 4.5) titration, field (mg/L as calcium carbonate) (00410)	Bicarbonate, water, unfiltered, fixed endpoint (pH 4.5) titration, field (mg/L) (00440)	Carbonate, water, unfiltered, fixed endpoint (pH 8.3) titration, field (mg/L) (00445)
004S004E14Q001S	334905116320901	07/29/1974	–	267	8.3	90	110	0
		09/13/1973	20	272	7.9	91	111	0
		05/15/1973	29	259	8	92	112	0
		07/24/1972	25.6	315	7.9	108	132	0
		08/24/2006	26.3	316	8	–	–	–
004S004E15J001S	334920116324701	04/30/1951	–	481	–	220	268	0
		08/14/1968	–	570	7.1	195	238	0
004S004E23E001S	334843116322601	11/22/1978	28	495	7.2	190	–	–
		06/15/1954	15.6	135	6.9	62	76	0
		10/25/1968	17.8	208	7.9	72	88	0
		09/12/1973	20	222	8.1	77	94	0

State well no.	USGS ID	Date (mm/dd/yyyy)	Calcium, water, filtered (mg/L) (00915)	Magnesium, water, filtered (mg/L) (00925)	Sodium, water, filtered (mg/L) (00930)	Sodium fraction of cations, water, percent in equivalents of major cations (00932)	Potassium, water, filtered (mg/L) (00935)	Chloride, water, filtered (mg/L) (00940)
004S004E14Q001S	334905116320901	07/29/1974	20	1.2	38	58	3.1	12
		09/13/1973	18	0.7	41	64	2	15
		05/15/1973	15	2.4	40	64	2	13
		07/24/1972	22	1.7	42	58	3.3	18
		08/24/2006	28.1	1.44	32.5	47	3.56	16.1
004S004E15J001S	334920116324701	04/30/1951	43	14	47	–	–	11
		08/14/1968	66	3	44	34	6	21
004S004E23E001S	334843116322601	11/22/1978	59	3.4	44	36	5.5	15
		06/15/1954	15	3	8	25	2.3	4
		10/25/1968	20	5	13	28	3	9
		09/12/1973	23	3	31	48	2	12

Table 7. Historical and recent water-quality data from three production wells near the Agua Caliente Spring, California.—Continued

[See figure 1 for site locations. State well no., see well-numbering diagram in text; the five-digit USGS parameter code in parentheses is used to uniquely identify each water constituent or property in the NWIS database. **Abbreviations:** USGS ID, U.S. Geological Survey identification number which is the unique number for each site in the USGS NWIS (National Water Information System) database; mm/dd/yyyy, month/day/year; °C, degrees Celsius; µS/cm, microsiemens per centimeter at 25°C; mg/L, milligrams per liter; µg/L, micrograms per liter; ft, feet; ft BLS, feet below land surface datum; <, less than; M, presence verified but not quantified; -, no data]

State well no.	USGS ID	Date (mm/dd/yyyy)	Sulfate, water, filtered (mg/L) (00945)	Fluoride, water, filtered (mg/L) (00950)	Silica, water, filtered (mg/L) (00955)	Boron, water, filtered (µg/L) (01020)	Iron, water, filtered (µg/L) (01046)	Aluminum, water, filtered (µg/L) (01106)
004S004E14Q001S	334905116320901	07/29/1974	19	0.5	-	<20	<10	-
		09/13/1973	15	0.9	-	100	-	-
		05/15/1973	16	0.6	-	<20	-	-
		07/24/1972	18	0.5	-	50	20	-
		08/24/2006	25.3	0.28	20.4	27	<6	3.2
004S004E15J001S	334920116324701	04/30/1951	32	0.4	12	40	M	M
		08/14/1968	48	0.2	-	-	-	-
		11/22/1978	44	0.2	27	110	20	-
004S004E23E001S	334843116322601	06/15/1954	3	0.2	-	60	-	-
		10/25/1968	9	1	-	0	-	-
		09/12/1973	18	<0.1	-	100	-	-

State well no.	USGS ID	Date (mm/dd/yyyy)	Residue, water, filtered, sum of constituents (mg/L) (70301)	Depth of hole (ft BLS) (72001)	Depth to top of sample interval (ft BLS) (72015)	Depth to bottom of sample interval (ft BLS) (72016)	Altitude of land surface (ft) (72000)
004S004E14Q001S	334905116320901	07/29/1974	153	980	622	958	425
		09/13/1973	147	980	622	958	425
		05/15/1973	144	980	622	958	425
		07/24/1972	170	980	622	958	425
		08/24/2006	193	-	-	-	425
004S004E15J001S	334920116324701	04/30/1951	291	438	164	435	453
		08/14/1968	305	438	164	435	453
		11/22/1978	312	-	-	-	-
004S004E23E001S	334843116322601	06/15/1954	73	488	240	472	435
		10/25/1968	103	488	240	472	435
		09/12/1973	135	488	240	472	-

A

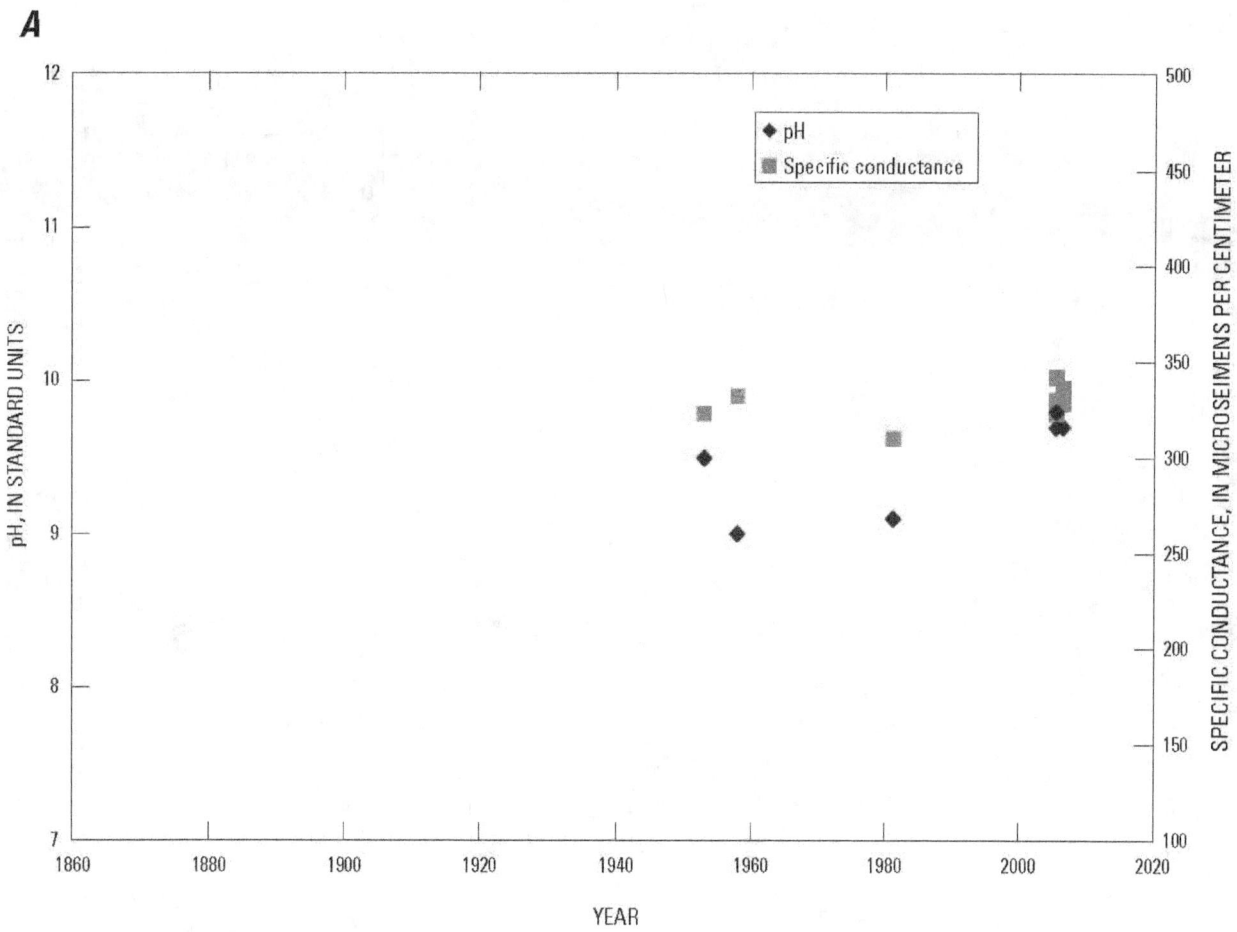

Figure 23.   Historical (*A*) pH and specific conductance and (*B*) sodium and chloride concentration in water sampled from the Agua Caliente Spring, California, 1876–2006.

B

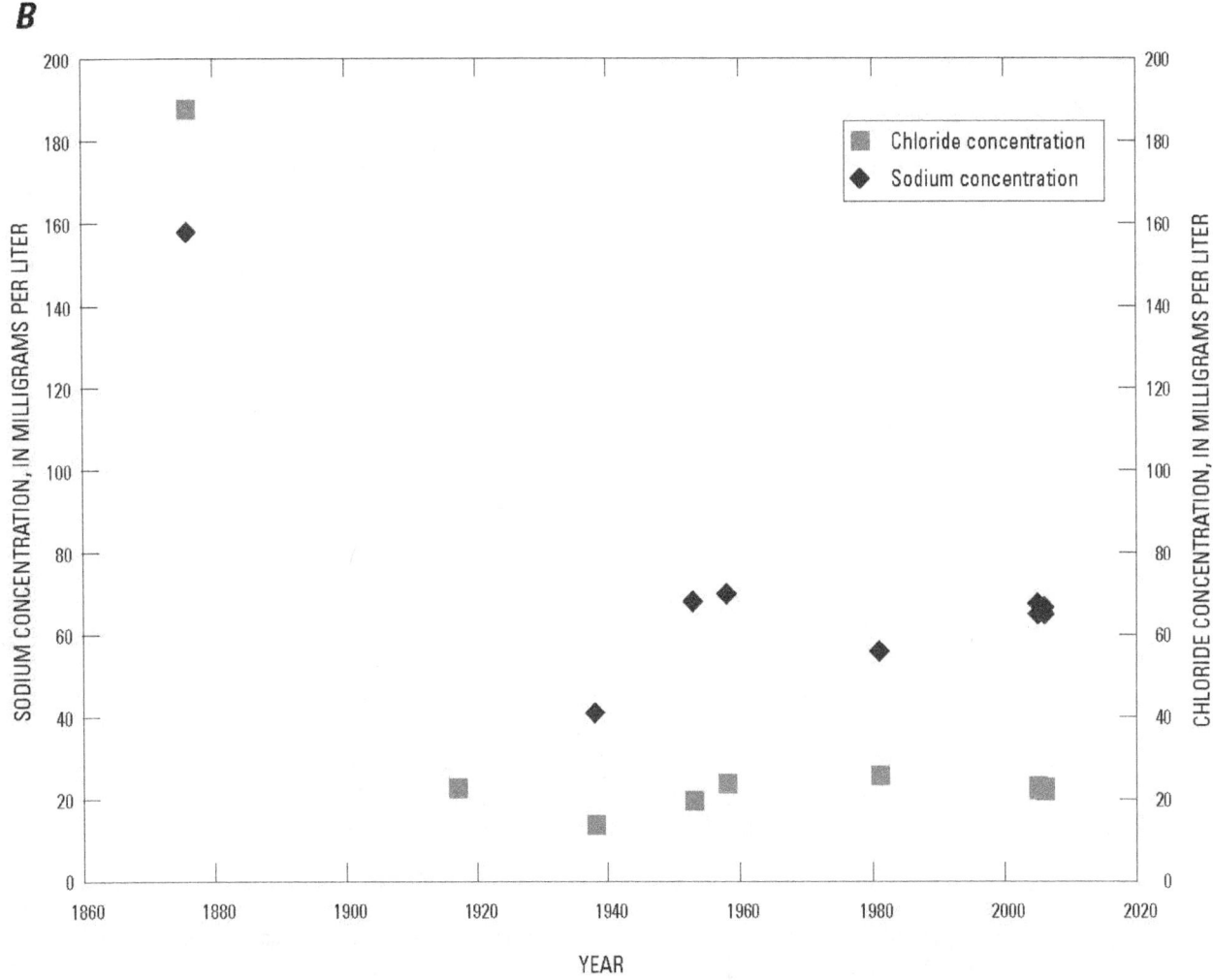

Figure 23.—Continued

Electrical conductance is caused by ions dissolved in the water; therefore, specific conductance (electrical conductance at 25°C) serves as a surrogate for residue-on-evaporation (dissolved-solids) and sum of dissolved constituents concentration. The ratio of specific conductance, in microsiemens-per centimeter (µS/cm), to residue-on-evaporation (ROE) concentration, in mg/L, generally ranges between about two-thirds and three-quarters, depending on relative contribution of the major ions. Both are general indicators for water quality, and are used in this study to compare samples collected for this study with historical data collected from Agua Caliente Spring. The specific-conductance measurements of samples collected from springs for this study range from about 200 µS/cm at Chino Warm Spring to about 600 µS/cm at Fenced Spring (fig. 24).

Inspection of the historical specific-conductance data collected from Agua Caliente Spring indicates little or no change in specific conductance since measurements were first recorded in 1953 (fig. 23A).

Dissolved oxygen and hydrogen sulfide provide an indication of the redox (oxidation-reduction) state of the water. Oxygen from the atmosphere is contained in the water at the time of recharge, and subsequently is consumed by biological (respiration and microbial) activity in the subsurface, resulting in a dissolved oxygen concentration in groundwater (and water as it issues from the spring) that is undersaturated with respect to water that is in equilibrium with oxygen in the atmosphere (table 6). Hydrogen sulfide results from microbial sulfate reduction after all the available oxygen has been consumed.

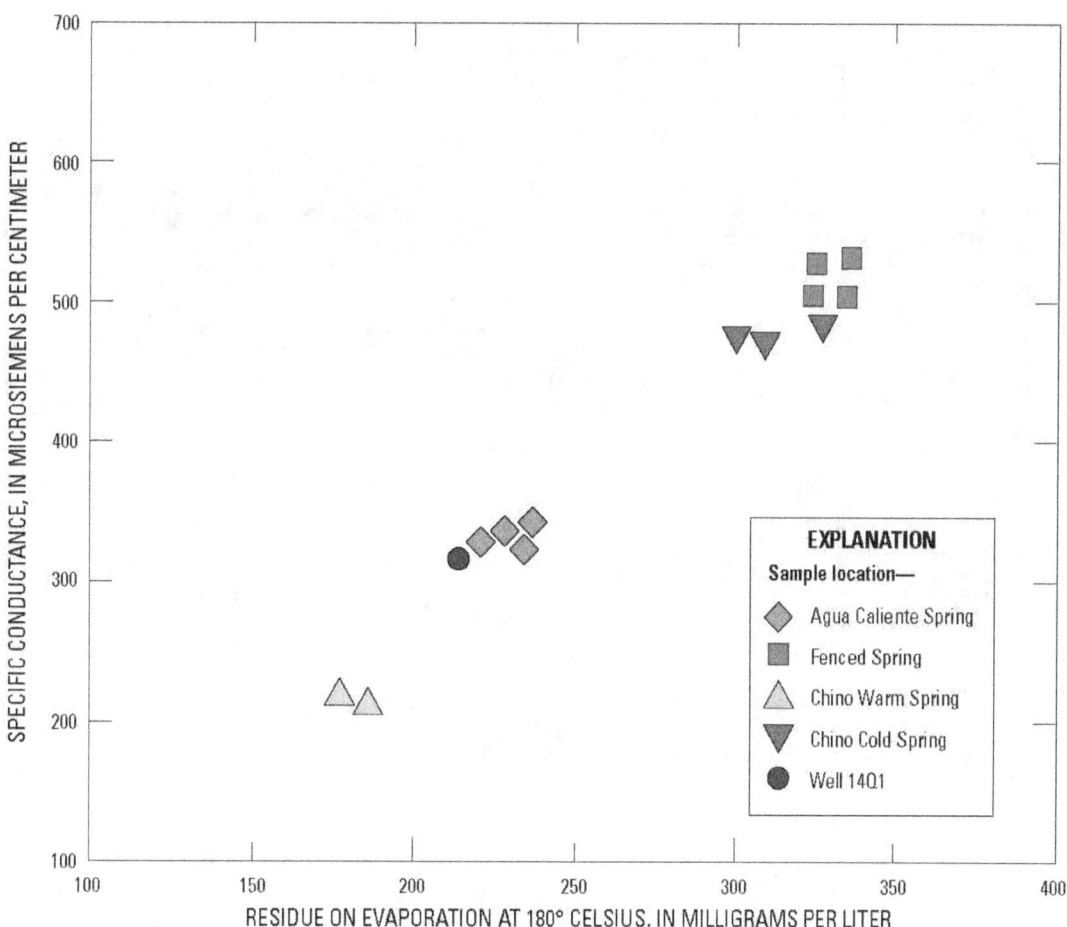

Figure 24. Relationship between specific conductance and residue-on-evaporation at 180 degrees Celsius in water from the Agua Caliente Spring area, California, 2005–06.

At thermodynamic equilibrium, dissolved oxygen and hydrogen sulfide should not coexist; however, a small amount of dissolved oxygen was found in water from all three thermal springs sampled for this study (Agua Caliente, Fenced, and Chino Springs; table 6). Samples measured immediately in the field, and even after several days in the sealed bottles, contained dissolved oxygen despite the presence of hydrogen sulfide (H_2S) and methane (CH_4) and the absence of nitrate (NO_3; tables 6 and 8). The coexistence of such reduced and oxidized constituents suggests some exchange must be taking place with oxic soil gases during upward transit of the thermal water through the unsaturated zone.

Dissolved CO_2 concentration in a water sample from the Chino Cold Spring was about 20 mg/L (table 8), which

is to be expected for groundwater with near-neutral pH. This is in marked contrast to all three thermal springs, where the CO_2 concentration only is about 0.1 mg/L, or less, because it has been "scrubbed" (converted to dissolved bicarbonate and carbonate) by the spring water's high pH (greater than 8.8). Analysis of captured gas bubbles collected from Agua Caliente Spring confirmed this result, in that they have even less CO_2 than air itself (table 9). This result contrasts markedly with samples of most groundwater, which typically have CO_2 partial pressures greater than the atmosphere at the land surface. Note that O_2 was detected at low concentrations in the gas bubbles, just as it was in the dissolved phase, which supports the field measurements of dissolved gases.

Table 8. Common atmospheric gas concentrations in springs sampled for the Agua Caliente Spring study, California, 2005–06.

[See figure 1 for site location. **Abbreviations:** USGS ID, U.S. Geological Survey identification number which is the unique number for each site in the USGS NWIS (National Water Information System) database; mm/dd/yyyy, month/day/year; hh:mm, hour:minute; °C, degrees Celsius; mg/L, milligrams per liter; ccSTP/L, cubic centimeter gas at standard temperature (25°C) and pressure (1 bar) per liter of water]

Local identifier	USGS ID	Date (mm/dd/yyyy)	Time (hh:mm)	Methane (mg/L)	Methane, precision, ±1-sigma[1]	Carbon dioxide (mg/L)	Carbon dioxide, precision, ±1-sigma[1]	Nitrogen (mg/L)	Nitrogen, precision, ±1-sigma[1]
Agua Caliente Spring	334924116324301	04/27/2005	08:00	0.0222	0.0007	0.0000	0.0000	16.5486	0.1527
		09/09/2005	09:00	0.0228	0.0002	0.0083	0.0019	17.5475	0.0197
		04/05/2006	10:00	0.0307	0.0004	0.0000	0.0000	17.0426	0.0498
		09/11/2006	10:00	0.0329	0.0004	0.0000	0.0000	18.1909	0.1270
Fenced Spring	334401116321201	04/29/2005	10:00	0.0326	0.0003	0.1223	0.0030	13.7977	0.0234
		09/08/2005	09:45	0.0181	0.0004	0.0942	0.0005	13.7468	0.0172
		04/04/2006	10:30	0.0390	0.0110	0.1169	0.0086	13.9041	0.1739
		09/12/2006	10:15	0.0657	0.0085	0.1318	0.0128	14.3014	0.3793
Chino Warm Spring[3]	335030116360401	04/28/2005	09:00	0.0153	0.0006	0.0000	0.0000	11.6456	0.1212
		09/10/2005	10:00	0.0108	0.0002	0.0056	0.0023	11.1861	0.0402
Chino Cold Spring	335030116360601	09/12/2006	10:15	0.0000	0.0000	19.9317	0.2719	14.8514	0.0358

Local identifier	USGS ID	Date (mm/dd/yyyy)	Oxygen (mg/L)	Oxygen. Precison, ±1-sigma[1]	Argon (mg/L)	Argon precision, ±1-sigma[1]	Argon[2] (ccSTP/L)	Water temperature (°C)	Excess air (mg/L)	Recharge temperature (°C)
Agua Caliente Spring	334924116324301	04/27/2005	0.1535	0.0085	0.5339	0.0019	0.2996	41.2	4.3	15.4
		09/09/2005	0.1390	0.0013	0.5553	0.0001	0.3116	41.0	5.7	14.6
		04/05/2006	0.1820	0.0197	0.5330	0.0010	0.2991	41.0	5.1	17.0
		09/11/2006	0.1802	0.0022	0.5560	0.0026	0.3120	40.0	6.1	16.3
Fenced Spring	334401116321201	04/29/2005	0.1518	0.0014	0.5018	0.0020	0.2816	31.2	0.8	15.2
		09/08/2005	0.1267	0.0017	0.4971	0.0012	0.2789	33.0	0.9	15.7
		04/04/2006	0.3092	0.1633	0.4924	0.0040	0.2763	32.0	1.4	16.9
		09/12/2006	0.1735	0.0034	0.4984	0.0074	0.2796	32.5	1.8	17.0
Chino Warm Spring[3]	335030116360401	04/28/2005	0.1173	0.0105	0.4322	0.0040	0.2425	40.5	0.2	18.2
		09/10/2005	0.1044	0.0015	0.4138	0.0012	0.2322	41.5	0.2	20.4
Chino Cold Spring	335030116360601	09/12/2006	0.2478	0.0583	0.5153	0.0011	0.2891	22.4	2.1	13.8

[1] Standard deviation (+ or − 1 sigma) is based on replicate samples.

[2] Calculated from argon concentration given in milligrams per liter.

[3] Gasses stripped from water.

Table 9.   Gas concentrations in bubbles isolated from water in the Agua Caliente Spring, California, April 5, 2006.

[See figure 2 for site location. **Abbreviations**: v/v in percent, indicated gas to total air volume ratio in percent; <, less than; –, no data]

Gas	Chemical symbol	Sample 1 (v/v in percent)	Sample 2 (v/v in percent)	Dry Air [1] (v/v in percent)
Helium	He	0.106	0.108	0.000524
Hydrogen	H_2	<0.0002	<0.0002	[2] 0.00007
Argon	Ar	1.009	1.022	0.934
Oxygen	O_2	0.460	0.355	20.9
Nitrogen	N_2	98.284	98.362	78.1
Methane	CH_4	0.134	0.138	[2] 0.00015
Carbon dioxide	CO_2	0.007	0.015	0.037
Ethane	C_2H_6	<0.0002	<0.0002	–
Hydrogen sulfide	H_2S	<0.0005	<0.0005	–
Carbon monoxide	CO	<0.001	<0.001	[2] 0.000015
Propane	C_3H_8	<0.0005	<0.0005	–
Butane	C_4H_{10}	<0.0005	<0.0005	–
Total, normalized		100.000	100.000	100

[1] Porcelli and others, 2002.

[2] Variable.

Major-Ion Composition

The major-ion composition of spring water is controlled by the natural chemistry of the recharge water and geochemical reactions, primarily dissolution and precipitation of minerals, in the subsurface. Reactions that occur include aluminosilicate dissolution, calcite precipitation, cation exchange on clays, and sulfate reduction. Groundwater from different sources can have different major-ion compositions as shown on the Piper diagram (fig. 25). A Piper diagram (Piper, 1944) displays the relative contribution of major cations and anions on a charge-equivalent basis, to the total ionic content of the water. Percentage scales show cation concentrations on a left triangle, and anion concentrations on a right triangle, and a central diamond integrates the data.

Calcium concentration for all samples displayed on the Piper diagram is controlled by the solubility of calcite,

as discussed in a later section of this report. The calcium composition is less than 5 percent of total cations (in mass units), and the sodium concentration composes greater than 95 percent of total cations in the Agua Caliente Spring and other thermal springs (Fenced and Chino Warm); whereas, sodium generally is about half, or less, in three wells (4S/4E -15J1, -23E1, and -14Q1) located within 1 mile of Agua Caliente Spring (data in tables 6 and 7, well locations on fig. 1). Another notable difference is that pH is nearly 10 (highly alkaline) in the Agua Caliente Spring but is only slightly higher than 7 (circum-neutral) in both regional groundwater from the valley floor and in the canyons (see, for example, data in table 6 for Chino Cold Spring). The large differences in sodium fraction and in pH are consistent with the absence of leakage or mixing from the regional aquifer into the thermal Agua Caliente Spring.

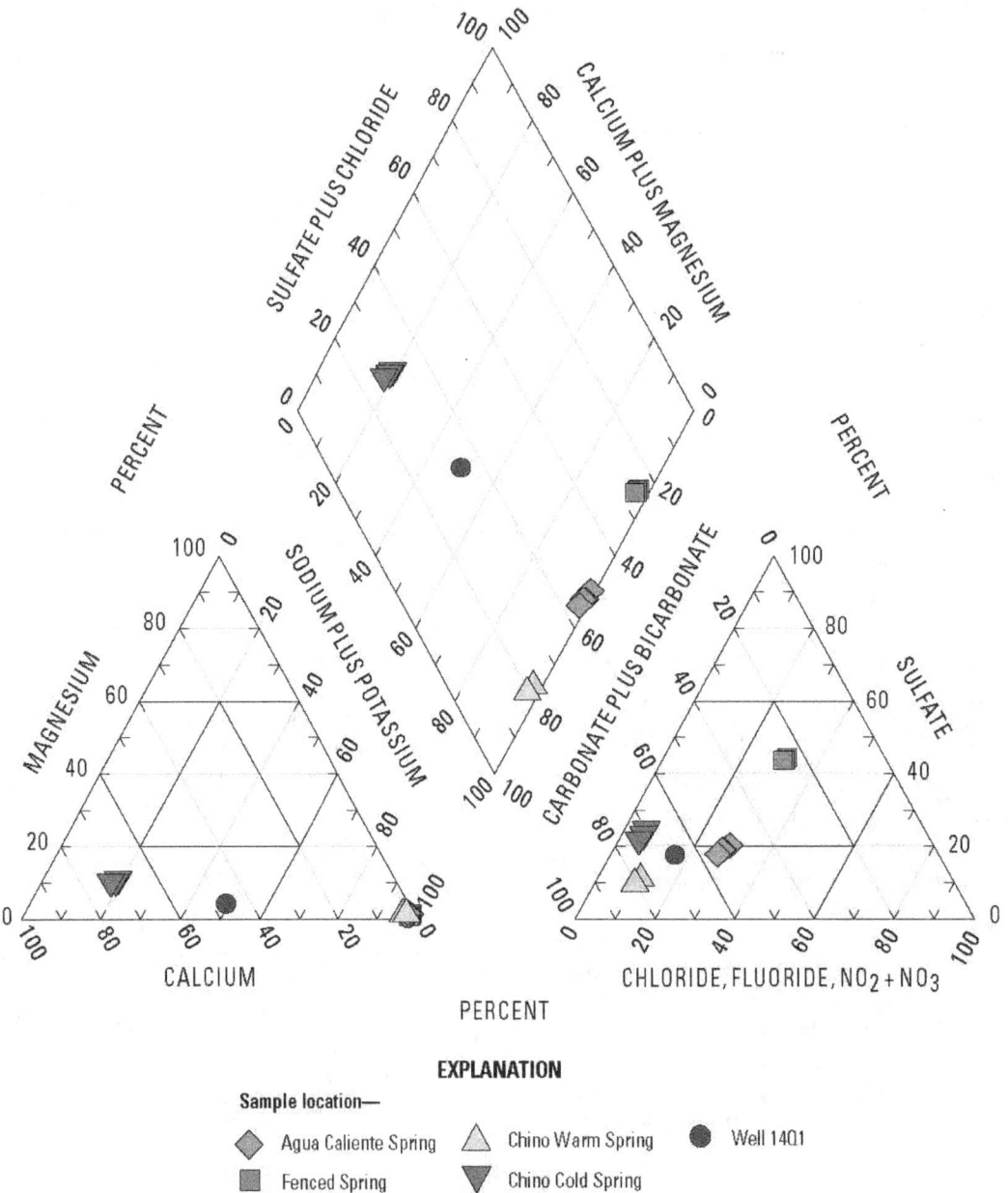

Figure 25.   Major-ion composition in water from the Agua Caliente Spring area, California, 2005–06.

Saturation Controls on Major-Ion Composition

The aqueous concentration of major ions is controlled by kinetics (rate of dissolution of minerals) and equilibrium (saturation with respect to minerals). A computerized geochemical model, PHREEQC (Parkhurst and Appelo, 1999), was used to ascertain whether the groundwater sampled in this study is in thermodynamic equilibrium (is saturated) with respect to any of a number of minerals, although this study was concerned primarily with calcite and silica.

Tests using PHREEQC show that solubility with respect to calcite controls the calcium concentration at both the thermal and non-thermal sites, and Fenced Spring is the only site that might be slightly undersaturated, with respect to calcite. Dissolved silica is supersaturated, relative to crystalline quartz and undersaturated relative to amorphous silica, which is common for groundwater; but, its concentration is very close to the solubility of chalcedony. These results could have implications for the possible cementation of the Agua Caliente Spring conduit (chimney) as described by Dutcher and Bader (1963).

Dutcher and Bader (1963) postulated that Agua Caliente Spring water enters the alluvial sediments from fractures in the underlying basement complex and then rises through the overlying valley-fill deposits through an uncemented conduit (chimney), which is confined laterally by poorly permeable material reworked and deposited by the spring. The test for solubility discussed previously was performed to examine the possibility that precipitation of minerals from the spring water could have produced a cemented conduit. Calcite and silica are plausible candidates as they often occur near the orifice of hot springs. Results using the computerized geochemical model PHREEQC indicate that the water at Agua Caliente Spring is at saturation (measured ion activity product is very close to the mineral's thermodynamic solubility product) with respect to both calcite and chalcedony. Reactions altering the composition of aluminosilicate minerals, especially feldspars, also provide a possible mechanism to provide cementation. However, it must be emphasized that while the model indicates the possibility of mineral deposition from solution on the basis of thermodynamic constraints, a definitive answer on whether it exists would require the direct examination of the material (from cores) itself. Because water-quality (major-ion) characteristics for Agua Caliente Spring are virtually identical to those in two shallow monitor wells, OW-1 and -2, several meters away from the spring (table 6), mineralogical examination of soil some distance away from the spring could confirm if there is cementation.

Temporal Trends

Specific conductance at the three thermal springs and one cool (ambient) spring sampled for this study had minimal change, either seasonally or annually, during 2005–06, indicating an absence of response to changes in discharge or precipitation (fig. 26). Major-ion composition in the Agua Caliente Spring also changed minimally (table 6), and the lack of temporal change suggests an old age for the water. In fact, long-term historical data indicate the water quality of Agua Caliente Spring itself has not changed appreciably during the last 100 years (table 4 and figs. 23A and B). The earliest sample obtained, in 1876, does appear somewhat more saline than all subsequent samples; however, from descriptions of the site, it is plausible that this could have been a grab sample from the marsh formed by the spring and, therefore, would have experienced evaporative concentration or dissolution of evaporated (efflorescent) salts from the margins of a marshy area.

Source of Anions

The chloride (Cl) to bromide (Br), chloride to sulfate, and chloride to boron ratios in groundwater samples were used in this study to help determine the source of anions to Agua Caliente Spring and other springs sampled for this study. Cl and Br are similar to one another in that both are chemically and biologically unreactive (they are conservative constituents). Salt in evaporating seawater is the source of both in precipitation. The Cl/Br mass ratio in seawater is about 285, so most natural waters have a ratio that is close to this value (Davis and others, 1998). The Cl concentration in atmospheric precipitation ranges from about 0.1 to almost 200 mg/L; concentrations decrease rapidly with increasing distance inland from the ocean. The Cl/Br ratio in atmospheric precipitation exhibits considerable scatter, but the trend also shows a decrease with increasing distance from the coast, from only slightly less than the ratio of 285 in seawater to about 100–200 in precipitation further inland, where the Cl concentration in precipitation has declined to less than a few milligrams per liter (Davis and others, 1998). The slight enrichment in Br relative to Cl in precipitation in coastal mountain ranges of southern California results in a Cl/Br ratio of about 200 in groundwater in coastal basins receiving mountain-front recharge (Schroeder and others, 1997; Land and others, 2004). Although both anions are highly soluble, Br is even more soluble than Cl, hence the insoluble residue (evaporite salt) that remains behind as the water evaporates is preferentially enriched in Cl, and the residual solution is preferentially enriched in Br. The net result is that the Cl/Br ratio in groundwater from an arid environment can be much higher than the ratio in local precipitation because halite in bedded salt deposits, or widely disseminated in the soil, has been re-dissolved. For example, by the time it reaches southern California, the Cl/Br ratio in Colorado River water is about four times higher than the ratio in seawater (Schroeder and others, 2002).

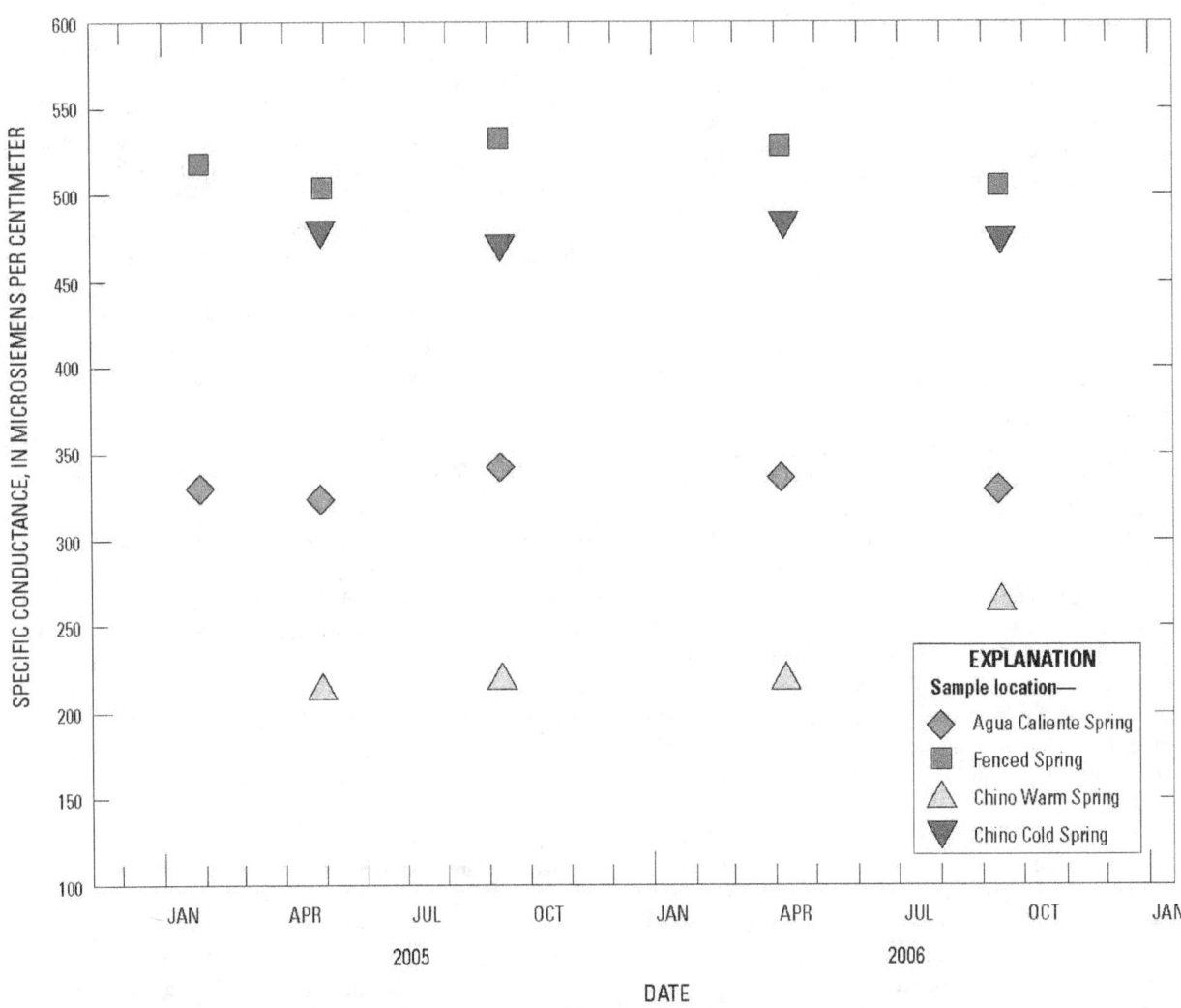

Figure 26. Specific conductance of water sampled from the Agua Caliente, Chino Warm, and Chino Cold Springs, California, 2005–06.

The chloride-to-bromide ratio in Agua Caliente Spring, Chino Warm Spring, and well 4S/4W-14Q1 is close to the ratio in seawater, and the chloride-to-bromide ratios in Fenced and Chino Cold Springs are slightly enriched (fig. 27A). This indicates that the origin of these highly soluble anions is salt from the ocean carried inland and deposited by precipitation, and that halides have no sources within the watershed, such as evaporite beds or broad dissemination within the soils. This finding contrasts with that of the less soluble boron

and sulfate, both of which are substantially enriched relative to seawater in samples from all four springs and the well, indicating that there are sources of boron and sulfate in the watershed other than direct input from precipitation (figs. 27B and C). Higher conductance and concentrations of both constituents in Fenced Spring indicate possible evaporative concentration at this spring (table 6). Although boron and sulfate concentrations are enriched, they still are substantially unsaturated in respect to common source minerals.

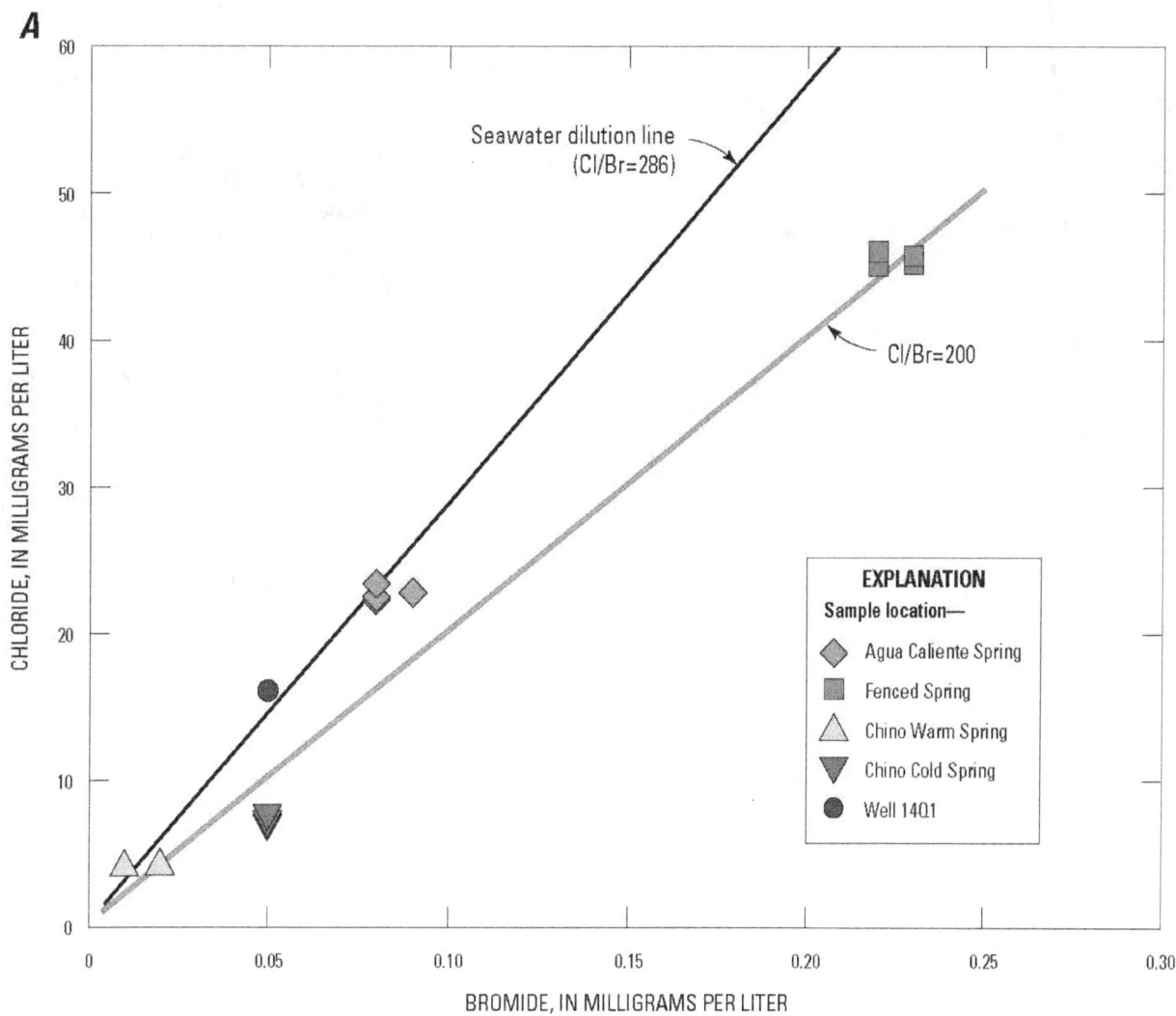

Figure 27. Relationship between chloride and (A) bromide, (B) sulfate, and (C) boron in water sampled in the Agua Caliente Spring area, California, 2005–06.

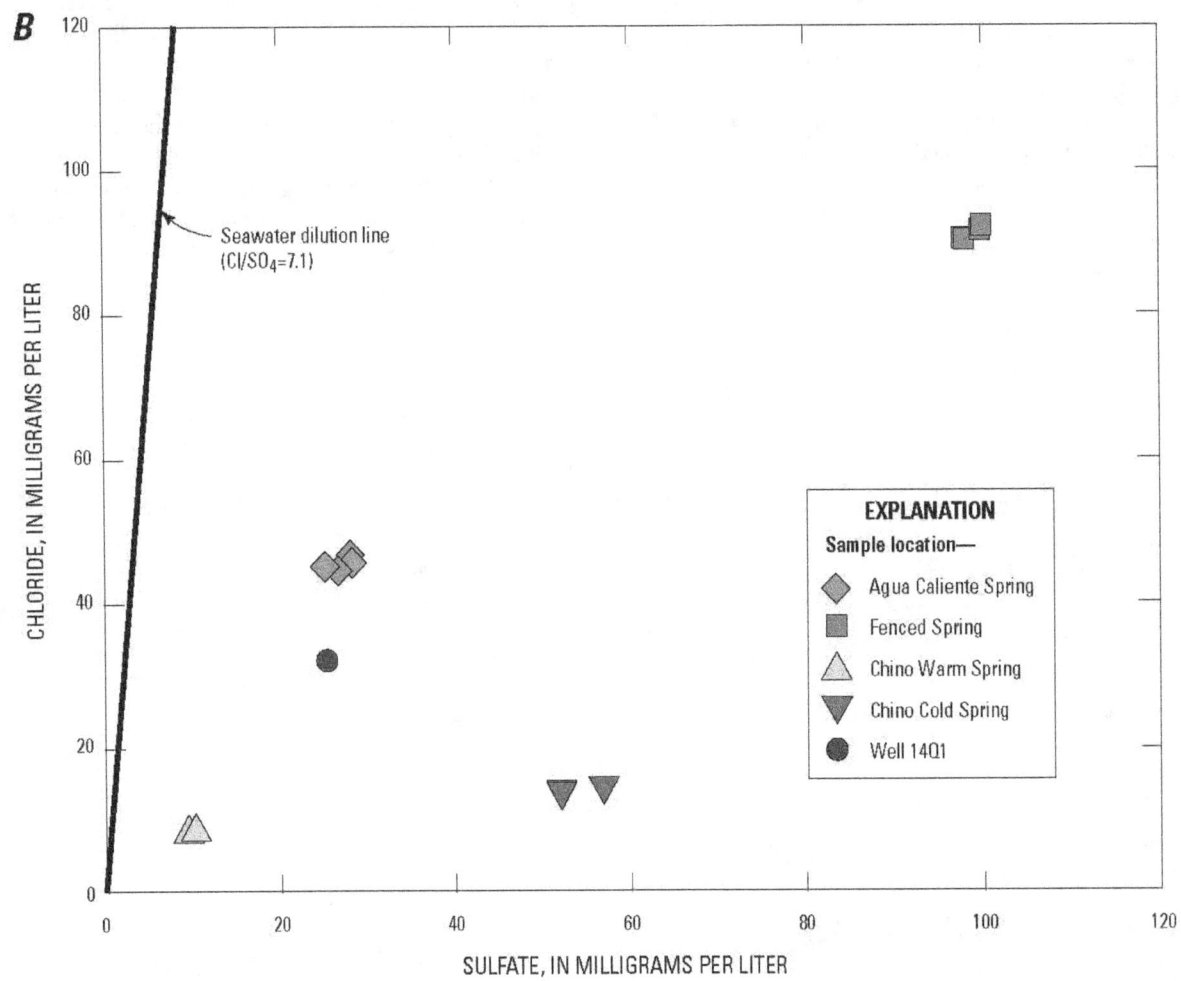

Figure 27.—Continued

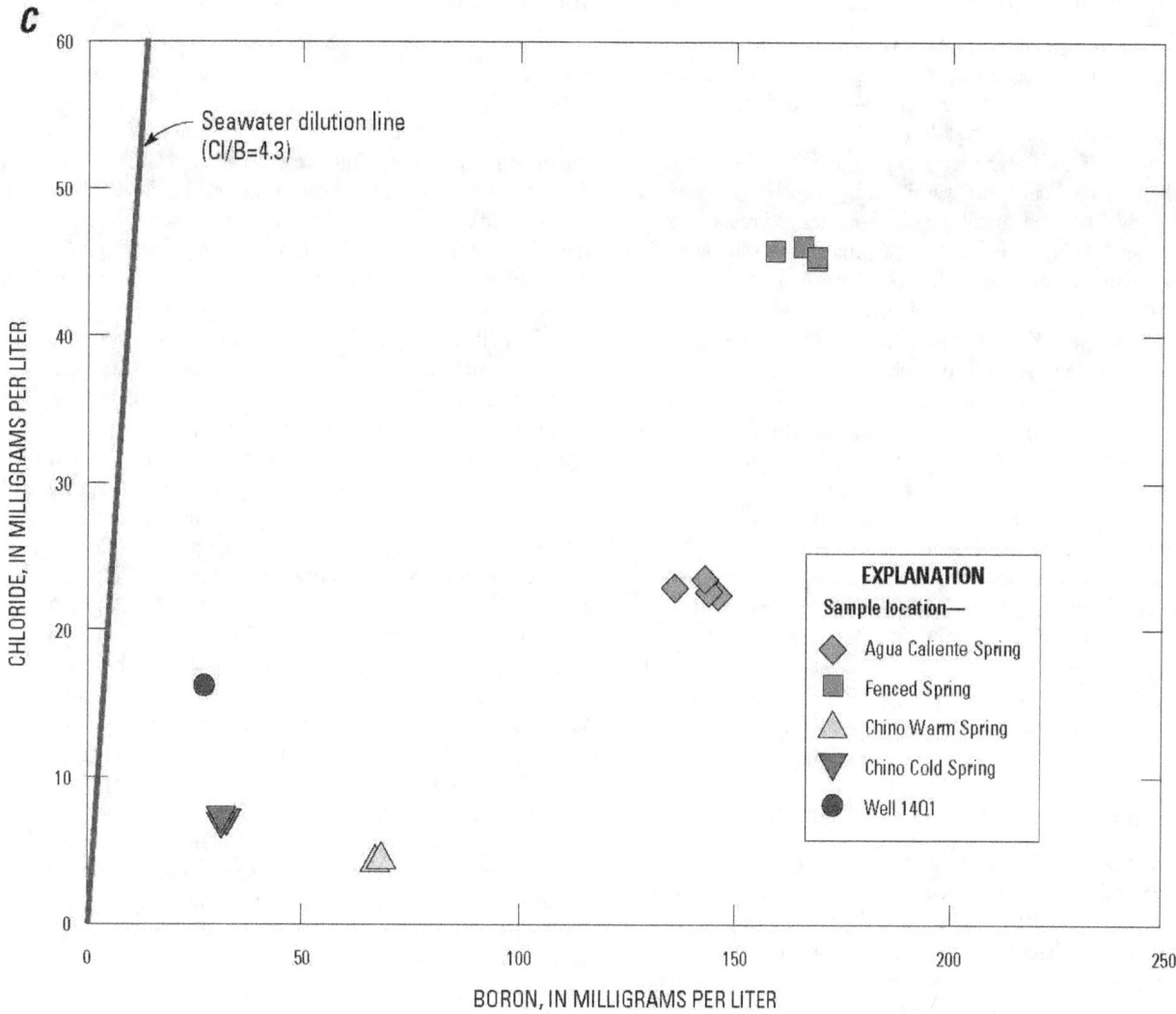

Figure 27.—Continued

Temperature of the Geothermal Source Water

Temperature of the geothermal reservoir (geothermal source water) was estimated by using two equations that are based on aqueous concentrations—the solubility of chalcedony (the cryptocrystalline form of silica) and an empirical relationship between Na, K, and Ca believed most likely to be applicable to "warm" springs, such as those in this study. These chemical geothermometers correspond to equations (a) and (g) in table 2 of Corbaley and Oquita (1986) and equations 2 and 5 in table 7 of Kharaka and Mariner (1989), respectively. The computerized geochemical model, SOLMINEQ.88 (Kharaka and others, 1988), was used to estimate geothermal reservoir temperatures for the three thermal springs in this study.

Using SOLMINEQ.88 (Kharaka and others, 1988) with Na, K, and Ca concentrations measured in samples from Agua Caliente, Fenced, and Chino Warm Springs (table 6) yields reservoir temperatures ranging from 61 to 71°C. This computed temperature range compares reasonably well with temperature calculations based on aqueous equilibration with chalcedony, which yields estimates ranging from about 50 to 60°C for Fenced Spring, about 70 to 80°C for Chino Warm Spring, and about 60 to 70°C for Agua Caliente Spring. Both methods confirm a moderate temperature, far below the boiling point of water, for the geothermal source water at all three warm springs.

Nutrients

Nitrogen and phosphorus concentrations are expected to be low at the springs sampled for this study because the water source is located far from human influences. Therefore, one would expect "background" nitrogen concentrations of about 2 mg/L, and that all the nitrogen at the spring's source would be in the form of nitrate if oxygen is present, which is the case in the spring samples collected for this study (table 6). In the samples from the thermal springs (Agua Caliente, Fenced, and Chino Warm Springs), where oxygen is absent, and hydrogen sulfide is present, microbial denitrification has converted the nitrate to nitrogen gas, and little if any ammonia is present (less than 0.04 mg/L; table 6). Plants (algae) in streams or pools that receive water from springs or seeps will take up any inorganic nitrogen (nitrate or ammonia) that is available and convert it to organic nitrogen. Phosphorus concentrations are very low (less than 0.02 mg/L as orthophosphate; table 6) because there is no significant source and because soils are effective at adsorbing phosphate.

Trace Elements

Most trace elements are present at very low concentrations in water samples collected from springs for this study (table 6). However, several trace elements do exhibit rather large differences in concentration among the three thermal springs (Agua Caliente, Fenced, and Chino Warm Springs) and between the regional groundwater sampled from Chino Cold Spring and well 4S/4E-14Q1, including barium (Ba), boron (B), lithium (Li), tungsten (W), vanadium (V), and uranium (U), and strontium (Sr) and its stable isotopes (table 6).

Ba concentration is near 1 µg/L in water from the 3 thermal springs and about 50 times higher in Chino Cold Spring and well 4S/4E-14Q1. B concentration is about two times higher in Chino Warm Spring than in Chino Cold Spring and well 4S/4E-14Q1, and about two times higher in the other two thermal springs than in Chino Warm Spring. Li concentration is a little more than two times higher in Agua Caliente Spring compared to Fenced Spring, and a little more than four times higher in Fenced Spring than in Chino Warm Spring. W concentration ranges from a low of 0.2 µg/L at Chino Cold Spring to more than 100 times higher at Agua Caliente Spring, and differs among all 5 sites (3 thermal and 2 non-thermal) sampled in this study. V concentration is less than 0.1 µg/L in all three thermal springs, compared to about 5 µg/L at Chino Cold Spring and about 20 µg/L in well 4S/4E-14Q1. U concentration is less than 0.04 µg/L in all three thermal springs, compared to about 4 µg/L in well 4S/4E-14Q1 and about 20 µg/L at Chino Cold Spring. Sr displays a pattern that is similar to other trace elements insofar as its concentration is about 30 times higher at the 2 non-thermal sites (well 4S/4E-14Q1 and Chino Cold Spring) compared to the 3 thermal sampling sites (Agua Caliente Spring, Fenced Spring, and Chino Warm Spring; table 6). Such large differences in the trace-element concentrations between the thermal springs and the regional groundwater lend additional support to the earlier inference, made on the basis of major-ion water quality, that no mixing occurs between the thermal water and regional aquifer through which it rises at either Agua Caliente or Chino Warm Spring. In addition, the large differences in concentration observed for several trace elements among the three thermal springs indicate that the geothermal source for each warm spring must be of limited areal extent (is localized) as opposed to regional.

In contrast to Sr, for which the concentration is the same at all three thermal springs (within analytical precision), the $^{87}Sr/^{86}Sr$ isotope signatures are significantly different in samples from each of the springs. Strontium isotopes are measured and reported directly as atom ratios ($^{87}Sr/^{86}Sr$) to four or five significant digits. ^{87}Sr in rocks and minerals is produced by decay of the radioactive isotope of rubidium, ^{87}Rb, which has a half-life of 5×10^{10} years (Aldrich and others, 1956). Strontium ratios depend on both age and rubidium content of rocks and minerals; hence, there can be large differences related both to composition and age (Turekian and Wedepohl, 1961). The dissolution and exchange of strontium between solid and aqueous phases is fairly rapid, so groundwater acquires a strontium isotopic signature that matches the deposit with which it is in contact. Variations in strontium isotopes due to differences in time of sample collection are used in this study to assess whether water that recharges the spring flows through the same deposits at all times, or whether the relative contribution from different deposits (formations) changes seasonally and annually in response to changes in rainfall amount. Strontium isotopes provide an independent method of assessment that can corroborate conclusions inferred from major-ion composition because each approach is affected by different underlying factors, implying different geologic sources of Sr in the springs; however, the actual source rock cannot be determined because the strontium isotope ratios for the geologic sources are unknown.

The $^{87}Sr/^{86}Sr$ isotope signatures are different from the third significant digit after the decimal point in samples from the three thermal springs (table 6), indicating that the geologic deposits that compose the thermal reservoir for each of the springs are different. Repeat samples from the thermal springs had the same $^{87}Sr/^{86}Sr$ isotope signature to the fourth significant digit after the decimal point, indicating that the geologic source of Sr did not change between sampling periods at the different springs. The $^{87}Sr/^{86}Sr$ isotope signatures of samples collected for this study support the previously mentioned inferences regarding the limited extent of the geothermal reservoir and the lack of mixing between the thermal springs and the regional groundwater.

Stable Isotopes of Oxygen and Hydrogen and Altitude of Recharge

Oxygen-18 (^{18}O) and deuterium (D, or ^2H) are naturally occurring stable isotopes of oxygen (O) and hydrogen (H). Analyses of data collected for these stable isotopes are used in this study to help determine the source and altitude of recharge.

Background

Isotopic ratios are expressed in delta notation (δ) as per mil (parts per thousand, ‰) differences relative to the standard known as Vienna Standard Mean Ocean Water (VSMOW) (Gonfiantini, 1978). The $\delta^{18}O$ ($^{18}O/^{16}O$ ratio) and δD ($^2H/^1H$ ratio) composition of precipitation throughout the world is linearly correlated because most of the world's precipitation is derived originally from the evaporation of seawater. This linear relationship is known as the global meteoric water line (Craig, 1961). Differences in the isotopic composition of precipitation occur along this line in response to trends with latitude and with the temperature of condensation. More negative values (depletion in the heavier relative to the lighter isotope) result when condensation takes place at colder temperatures and higher altitudes. The temperature effect, on the basis of measurements in North America and Europe, is -0.7 per mil per degree Celsius (‰/°C) for $\delta^{18}O$, which is equivalent to -5.6‰/°C for δD (Dansgaard, 1964). The altitude effect, on the basis of measurements taken on the western flank of the Sierra Mountain Range, is -2.3‰/km for $\delta^{18}O$ (Ingraham and Taylor, 1991; Rose and others, 1996). This is equivalent to -18.4‰/km (-0.56‰/100 ft) for δD if isotope data fall on the global MWL or on a local line that is parallel to the MWL. Water that has been partly evaporated is enriched in heavier isotopes relative to its original composition; these values plot to the right of the MWL (for δD as the vertical and $\delta^{18}O$ as the horizontal axis).

Isotopic Composition of Groundwater Samples

The $\delta^{18}O$ and δD composition of groundwater samples collected from the study area ranged from -9.07 to -11.66 and -66.3 to -81.1 per mil, respectively (fig. 28; table 6). Water samples for analysis of stable isotopes of oxygen and hydrogen were obtained from four additional sites where no other water-quality data were obtained—three springs in Palm Canyon (Dos Palmas, Indian, and Cedar) and Chino Canyon Creek (fig. 1). Chino Canyon Creek drains the Chino Canyon where it flows beneath the Palm Springs Aerial Tramway from Mt. San Jacinto (summit at 10,834 ft).

Seven years of six-month seasonal (summer and winter) monitoring, from 1982 to 1989, reveal a weighted mean δD of -84‰ in precipitation from a fixed station at an altitude of 8,330 ft (2,450 m) on Mt. San Jacinto (Friedman and others, 1992; Smith and others, 1992; and Gleason and others, 1994).

Both δD and δ¹⁸O data were obtained for only 2 of the 7 years, 1986 and 1987, and the weighted means for this short duration are –77.2‰ and –10.96‰, respectively (calculated from data in Friedman and others, 1992). These data would plot very nearly on the global MWL; therefore, it is reasonable to impute a value of –11.75‰ for δ¹⁸O if oxygen isotopes

had been analyzed for the entire 7-year period of precipitation monitoring. The δD in samples from all sites for this study is less negative (heavier) than this 7-year average in precipitation (fig. 28), as is to be expected because all are from altitudes that are considerably lower than the Mt. San Jacinto station.

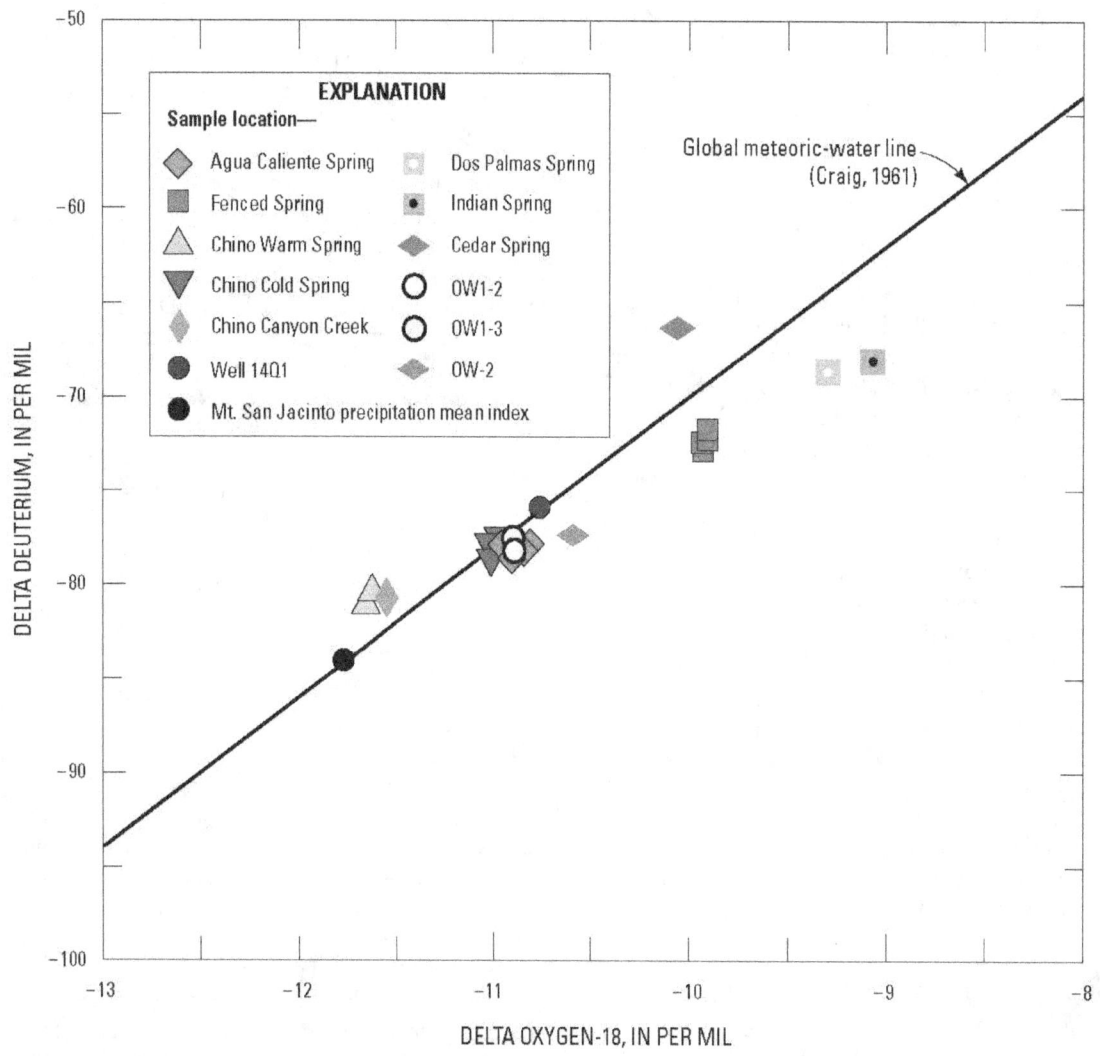

Figure 28.   Relationship between delta deuterium and delta oxygen-18 in water sampled from the Agua Caliente Spring study area, California, 2005–06.

The isotope data from most of the sites sampled for this study fall on, or near, the global MWL (fig. 28). Those that lie to the right of the MWL probably have undergone a small amount of evaporation. These sites include Fenced, Indian, and Dos Palmas Springs, and the three shallow observation wells (OW1-2, OW1-3, and OW-2) near the Agua Caliente Spring. The finding of some evaporation at these shallow observation wells is not surprising given their location within a shallow perching zone in an arid climate, sustained only by lateral flow from the spring. The possibility of evaporative concentration at Fenced Spring was suggested as well on the basis of its comparatively higher salinity, sulfate, and boron concentrations than in the other two thermal springs.

The datum point for Cedar Spring (δD is $-66.3‰$, $\delta^{18}O$ is $-10.06‰$) is unusual insofar as it lies so far to the left of the MWL, and isotopic data from springs almost always lie on, or to the right, of the MWL. This suggests the possibility of analytical error because temperatures are not high enough for water-rock interactions to alter the water isotopes. If the error were in the measurement of one isotope ratio only, and the correct value could be estimated by extrapolating either vertically or horizontally from the reported data point to intercept the MWL, the corrected value would be either δD = $-70.5‰$ or $\delta^{18}O$ = $-10.06‰$, respectively. Cedar Spring is located at an altitude of 6,433 ft, and such a high altitude favors adjustment to the more negative value for δD. Of course, collection and analysis of another sample from Cedar Spring is the only certain method to resolve the uncertainty.

Altitude of Recharge

The source of water for the Agua Caliente Spring, and other spring and wells sampled for this study, is believed to be the infiltration of precipitation in the San Jacinto Mountains. Infiltration takes place over a range of altitudes in the mountains, but water isotopes can be used to calculate the "average" altitude at which recharge occurs for several samples whose data fall on or near the MWL. The method extrapolates from the isotopic composition of precipitation at the Mt. San Jacinto station, by using the altitude effect of isotope ratios measured on the western flank of the Sierra Mountain Range (Ingraham and Taylor, 1991; Rose and others, 1996), presented earlier in this section of this report, to calculate the average altitude of recharge for a sample. The results that follow were obtained by substituting δD values of the spring and well samples (means if a site was sampled more than once) in the following equation. An analogous calculation also could be done using $\delta^{18}O$ values of the spring and well samples, and it would yield an identical result.

$$E = E_{SJP} + \left(\frac{\delta_{s,w} - \delta_{SJP}}{Z} \right), \qquad (1)$$

where

E is the caluclated evaluation of recharge to a spring or well,

E_{SJP} is the elevation of the Mt. San Jacinto precipitation (8.330 feet),

$\delta_{s,w}$ is the hydrogen-isotope ratio for the spring (s) or well (w),

δ_{SJP} is the hydrogen-isotope ratio for the Mt. San Jacinto precipitation station (-84 per mil), and

Z is the effect of altitude on isotopic compositon of precipitation (assumed to be -0.56 per mil/ 100 feet, as discussed earlier).

The altitude of recharge was calculated to be about 6,880 ft for well 4S/4E-14Q1; 7,740 ft for Chino Canyon Creek; 7,260 ft for Chino Cold Spring; 7,750 ft for Chino Warm Spring; and 7,290 ft for Agua Caliente Spring. The calculation yields a recharge altitude of about 6,100 for Fenced Spring; however, recharge probably was a few hundred feet higher because evaporation had caused the isotope ratios to become less negative (deviate to the right of the MWL) at this site. The calculated difference of 460 ft between Chino Warm Spring and Agua Caliente Spring may understate the actual difference in altitude of recharge for these two springs because water ages, described in the "Water Age" section of this report, indicate recharge to Agua Caliente Spring dates to near the end of the last North American glaciation, when it likely was colder and(or) wetter in the study area, which would cause isotope ratios to be lighter (more negative) than in postglacial (younger) water. The potential area of recharge estimated by this technique for each spring can be determined by comparing the calculated average altitude of recharge for an individual spring to the land-surface elevation contours shown on figure 1. The calculated average altitude of recharge for Agua Caliente Spring (7,290 ft) is higher than the highest elevation present in the Aqua Caliente Spring watershed (fig. 1), indicating that the source of water to the spring includes an area greater than the watershed boundary and(or) the isotopic ratio of precipitation was lighter (more negative) when the sample collected from the spring was recharged, which would result in overestimation of the altitude of recharge.

Dissolved Gases and Temperature of Recharge

In this study, the concentrations of atmospheric gases were used to estimate the temperature of recharge for the spring discharge sampled at Agua Caliente, Fenced, and Chino Warm Springs. Gas solubility is dependent on temperature, pressure, and salinity, with temperature dependence likely to exert the greatest influence on the spring's environment, followed by partial pressure (altitude). Solubility of all gases increases with decreasing temperature, and, therefore, measured concentrations can be used to calculate the temperature at which the infiltrating water was last in equilibrium with the atmosphere (shallow soil gas)—that is, the temperature at which recharge took place. The heavier (higher molecular weight) gases are more useful for estimating the temperature of recharge than the lighter ones because of their greater sensitivity to temperature change. For example, the solubility of neon (Ne) increases by only about 3 percent, while the solubility of argon (Ar) increases about 7 percent for each degree that temperature decreases (values are only approximate because temperature dependence is nonlinear). Partial pressure (and hence solubility) increases about 1.5 percent for each 100-m decrease in altitude between 3 km and sea level.

In addition to gas solubility changing with temperature, gas concentrations could also be affected by "excess" air. Excess air enters the groundwater as infiltrating water traps bubbles of air that subsequently dissolve at higher pressure as groundwater moves to greater depth beneath the water table. While this likely occurs to some extent in this study area, re-equilibration with the atmosphere as groundwater nears the land surface, or even at the time of sampling itself, probably is much more likely to compromise interpretation of the noble-gas data. Detailed discussions on the use of noble gases to determine recharge temperature can be found in Aeschbach-Hertig and others (1999), Aeschbach-Hertig and others (2000), and Manning and Solomon (2003). Ar and nitrogen (N_2) can be used to calculate recharge temperature as described in Heaton and Vogel (1981) in a manner analogous to that used for noble gases, although there is the added complication that N_2 can be produced by microbial denitrification of nitrate (NO_3) in the subsurface.

Calculated mean recharge temperatures using the concentration of Ar and N_2 are 15.8°C for Agua Caliente Spring, 16.2°C for Fenced Spring, 19.3°C for Chino Warm Spring, and 13.8°C for Chino Cold Spring (recharge temperatures calculated from individual samples given in

table 8). Any increase in the assumed altitude of recharge (decrease in partial pressure) will lead to a corresponding increase in concentration when adjusted to sea level, which translates to a lower calculated recharge temperature. On the basis of the stable-isotope evidence that showed the water in Fenced Spring could have experienced some evaporation during its history, it is plausible that the calculated recharge temperature at Fenced Spring is too high, although by no more than about 1°C. The calculated recharge temperature for Chino Warm Spring likely is too high because evidence indicates gasses were stripped at the spring discharge orifice. Low or negative "excess air" concentrations often suggest the gases have been "stripped" from a sample, and this appears to have happened at this spring, which is not surprising given the very high flow rate induced by forcing the entire discharge through a small-diameter pipe. Gas solubility decreases with increasing temperature; therefore, stripping of gases will yield a temperature estimate that is too high.

Analogous calculations based on noble gases yield mean recharge temperatures of 13.8°C for Agua Caliente Spring, 15.4°C for Fenced Spring, 18.7°C for Chino Warm Spring, 14.2°C for Chino Cold Spring, and 15.2°C for Well 14Q1 (recharge temperatures calculated from individual samples given in table 10). Comparison indicates recharge temperatures calculated by the two methods are within 2°C. With the exception of Chino Warm Spring, where reliable results could not be obtained because of gas stripping, the range in recharge temperatures for other sites is small, from about 14 to 16°C, as would be expected because the altitude of recharge calculated for each site differs by no more than about 1,000 ft.

Concentrations of Ar, the only gas analyzed by both the Reston CFC laboratory and LLNL, were found to be virtually identical (note the slight difference in reporting units when comparing concentrations between the two laboratories in tables 8 and 10), despite the very different methods used to collect water samples sent to each laboratory. This finding provides a reasonable level of confidence in the collection and analysis of dissolved gases for this study, although it does not resolve reservations regarding partial exchange of sample gases with gases in the shallow soil or atmosphere as the water nears or exits the spring at land surface. The calculated recharge temperature of approximately 14°C for Agua Caliente Spring is considered to be a maximum estimate given possible partial re-equilibration at a lower altitude (hence a warmer temperature) and stripping by gases present in its anaerobic thermal water.

Table 10.   Noble gas concentrations in springs and a well sampled for the Agua Caliente Spring, California, study, 2005–06.

[See figure 1 for site locations. State well No.: see well-numbering diagram in text. **Abbreviations:** USGS ID, U.S. Geological Survey identification number which is the unique number for each site in the USGS NWIS (National Water Information SYstem) database; mm/dd/yyyy, month/day/year; hh:mm, hour:minute; °C, degrees Celsius; ccSTP/g, cubic centimeter gas at standard temperature (25 °C) and pressure (1 bar) per gram of water; ccSTP/L, cubic centimeter gas at standard temperature (25 °C) and pressure (1 bar) per liter; STP, standard temperature (25 °C) and pressure (1 bar); –, no data]

Local identifier	USGS ID	Date (mm/dd/yyyy)	Time (hh:mm)	Helium-3/ Helium-4 (atom ratio) $\times 10^{-6}$	Helium-4 (ccSTP/g) $\times 10^{-7}$	Neon (ccSTP/g) $\times 10^{-7}$	Argon (ccSTP/g) $\times 10^{-3}$	Krypton (ccSTP/g) $\times 10^{-8}$	Xenon (ccSTP/g) $\times 10^{-8}$	Argon (ccSTP/L)[2]	Water tempera-ture (°C)	Recharge tempera-ture (°C)
Solubility at STP				[1] 1.4	0.441	1.78	0.284	6.18	0.854			
Agua Caliente Spring	334924116324301	04/27/2005	08:00	0.5430	37.8	73.5	1.342	54.8	4.61	0.2996	41.0	[3]
		09/09/2005	09:00	0.1300	65.2	2.08	0.3130	6.89	0.9360	0.3116	41.0	13.8
Fenced Spring	334401116321201	04/29/2005	10:00	0.1280	10.8	1.73	0.2870	6.39	0.8900	0.2816	31.2	15.0
		09/08/2005	09:45	0.1210	11.2	1.69	0.2740	6.18	0.8900	0.2789	33.0	15.9
Chino Warm Spring	335030116360401	04/28/2005	09:00	0.2780	3.94	1.52	0.2520	5.72	0.8310	0.2425	40.5	16.2
		09/10/2005	10:00	0.3660	1.89	1.44	0.2260	5.12	0.7030	0.2322	41.5	21.2
Chino Cold Spring	335030116360601	09/12/2006	10:15	0.8250	0.812	1.87	0.2900	6.37	0.8830	0.2891	22.4	14.2
Well 004S004E14Q001S	334905116320901	08/24/2006	10:00	0.4110	3.95	3.08	0.3900	8.03	0.9850	–	–	15.2

[1] Atmosheric ratio (Mamyrin and others, 1970).

[2] From analysis done by Reston CFC laboratory on common-gas samples (see table 8).

[3] Recharge temperature not calculated because noble-gas sample was contaminated by contact with air.

Radioactive Isotopes and Water Age

Age of the water, or the time elapsed since recharge, is arguably a spring's most important attribute when considering its susceptibility to climatic fluctuations and anthropogenic effects. The stability of water quality, as indicated by specific conductance at Agua Caliente Spring over nearly 100 years, suggests great age. Tritium (^{3}H) and carbon-14 (^{14}C), radioactive isotopes, were used in this study to determine water age, that is, the time it takes for recharging (infiltrating) water to move through the subsurface until it exits at the spring.

Background

^{3}H is a naturally occurring radioactive isotope produced by the interaction of high-energy cosmic rays with the upper atmosphere. For this study, its activity in samples was measured in tritium units (TU). One TU is equivalent to 1 tritium atom in 10^{18} hydrogen atoms (Taylor and Roether, 1982). Activity also often is reported in units of picocuries per liter (pCi/L). One TU is approximately 3.19 pCi. Both units are used in this report. Large amounts of ^{3}H were introduced into the environment during the atmospheric detonation of thermonuclear bombs beginning in the early 1950s. This raised the tritium concentration in precipitation several-hundred fold by the time testing was halted in 1963. ^{3}H levels in precipitation subsequently declined and are now near natural background levels in many locations. Activities ranging from 1.3 to 2.6 TU were measured in seven of eight rainfall samples collected during 2006 at one author's (R.A. Schroeder) residence at the coast in San Diego County. Activities in the Sierra Nevada snow pack were 3–4 TU during the winter of 2006–07 (R.L. Michel, USGS, written commun., 2007).

Absence of ^{3}H indicates water that is older than about 1950; very high activity indicates water that infiltrated near the time of the bomb peak in the early 1960s, and low activity suggests either very recent recharge or a mixture of water with different ages. The decay equation above can be used to calculate water age if ^{3}H activity at the time of infiltration is known; however, that is rarely the case. ^{3}H decays to the stable isotope of helium, helium-3 (^{3}He), so the initial activity of ^{3}H equals the sum of present-day ^{3}H plus ^{3}He (tritogenic helium). However, the direct application of this is complicated because ^{3}H decay is not the only source for ^{3}He. Additional details on the calculation of ^{3}H-^{3}He water ages can be found in Izbicki and others (2003).

Carbon 14 (^{14}C) is a naturally occurring radioactive isotope of carbon produced by cosmic-ray spallation with nitrogen in the upper atmosphere. Because of its long half-life, it can be used to date water that is much older than is possible with ^{3}H. ^{14}C is measured for dissolved inorganic carbon in water and reported as percent modern carbon (pmc) by comparing ^{14}C activities to the specific activity of National Bureau of Standards oxalic acid; 12.88 disintegrations per minute per gram of carbon in the year 1950 equals 100 pmc. Activity of ^{14}C varied in the past in response to fluctuations in its natural rate of production; it had been decreasing since the beginning of the Industrial Age 200 years ago from injection of CO_2 containing no ^{14}C ("dead" carbon) into the atmosphere from the burning of fossil fuel, but, activity increased with advent of the nuclear era. These factors are not important in the calculation of ^{14}C water ages for old water in this study because their effect is much less significant than is the exchange of aqueous dissolved inorganic carbon (DIC) with carbonate minerals. Carbonate minerals have little or no ^{14}C, which results in a decrease in ^{14}C activity and a corresponding increase in the apparent age of DIC.

Stable carbon isotopes [Carbon-13 and carbon-12], are used in this study in conjunction with radiogenic carbon-14, to make inferences about the extent to which this exchange could have caused the calculated ^{14}C age to overestimate the actual time elapsed since recharge. Carbon-13 (^{13}C) and carbon-12 (^{12}C) are naturally occurring stable isotopes of carbon. The ratio of these isotopes is expressed as differences relative to the standard Vienna Pee Dee Belemnite (VPDB ; Friedman and O'Neil, 1977; Coplen, 1994). Analytical precision is within 0.5‰ for δ^{13}C. Some representative δ^{13}C values are about 0‰ in carbonates, about –8‰ in atmospheric CO_2 at present, and range from about –20 to –28‰ in soil organic carbon. The isotopic compositions of DIC, carbonate, and bicarbonate in groundwater usually are derived from a mixture of atmospheric CO_2, oxidation (degradation or mineralization) of organic carbon, and dissolution of minerals. Because an exchange of carbon isotopes (equilibration) between carbonate minerals and DIC occurs, albeit slowly, groundwater could acquire less negative δ^{13}C values as it moves along a flow path from infiltration to exit at the spring.

Water Age

For the simplest assumptions (piston flow and no mixing), water age can be calculated from the radioactive decay equation:

$$A = A_0 e^{-\lambda t}, \text{ or its equivalent}$$
$$2.302(\log A - \log A_0) = -\lambda t \qquad (2)$$

where

A is the current activity,

A_0 is the activity in the water at its time of recharge, and

λ is the decay constant.

The half-life (time for activity of the nuclide to decay by half) is given by $t_{1/2} = 0.693/\lambda$. The half-life is 12.32 years for ^3H (Lucas and Unterweger, 2000) and 5,730 years for ^{14}C (Godwin, 1956).

Tritium is present at low concentrations, near present-day precipitation, in Chino Cold Spring and in the regional aquifer sample from well 4S/4E-14Q1 (table 6), indicating at least some contribution from water that is younger than 1950 (post-bomb). However, one of the samples from Chino Cold Spring and the sample from well 4S/4E-14Q1 have ^{14}C ages greater than 500 years before present, suggesting a mixture of recent and old groundwater (table 6). In contrast to the sample from well 4S/4E-14Q1, the samples from Agua Caliente Spring had ^3H at concentrations less than the measurement precision (less than 2 sigma), reinforcing the lack of mixing with groundwater in the regional aquifer that was indicated by chemical evidence presented earlier.

Carbon-14 activities for samples from Agua Caliente Spring, Chino Warm Spring, and Fenced Spring range from 16 to 43 percent modern carbon (table 6). Calculated ^{14}C age for three samples collected from Agua Caliente Spring range from about 14,200 to about 15,100 years before present, making its water the oldest of the three springs. This compares to somewhat younger ages of about 10,000 years before present for a single sample from Chino Warm Spring, and about 7,000 to about 8,300 years before present for three samples from Fenced Spring. Exchange between aqueous DIC and radiocarbon-dead carbonate in soils during movement of the groundwater will cause calculated ^{14}C ages to overstate the actual time since recharge; this effect can be estimated using $\delta^{13}C$ values in table 6. The $\delta^{13}C$ is about $-16‰$ in Fenced Spring, and a little less negative in Chino Cold Spring, where the presence of tritium (^3H) indicates a very young age. Hence, assuming the values for $\delta^{13}C$ represent the isotope ratio soon after the time of recharge to Agua Caliente and Chino Warm Spring, and that the isotope ratio at these two springs has been modified by exchange with soil carbonate, which is assumed to have a ratio of 0‰, there is a dilution of about 30 percent, which translates into a reduction of about 3,000 years in the water age for Agua Caliente Spring and Chino Warm Spring. This nonetheless indicates a very old age for water in both of these springs.

Simulation of Fluid and Heat Flow Near the Agua Caliente Spring

By Alan L. Flint and Peter Martin

Numerical models of fluid and heat flow were developed for the Agua Caliente Spring to (1) test the validity of the conceptual model presented by Dutcher and Bader (1963) that the Agua Caliente Spring enters the valley-fill deposits from fractures in the underlying basement complex and rises through more than 800 ft of valley-fill deposits through a washed-sand conduit and the surrounding low permeability deposits (spring chimney) of its own making, (2) evaluate whether water-level declines in the regional aquifer will influence the temperature of discharging water, and (3) determine possible sources of thermal water outside of the Agua Caliente Spring steel collector tank (possibilities are leaks in the casing and(or) the presence of a secondary orifice of the Agua Caliente Spring not contained by the steel collector tank). TOUGH2, an integrated finite-difference numerical code (Pruess and others, 1999), was used to develop a two-dimensional radial flow model and a three-dimensional flow model of fluid and heat flow near the spring. The radial flow model has shorter computer run times than the three-dimensional model, whereas the three-dimensional model provides more detailed simulations.

Model Discretization and Boundary Conditions

The axisymmetric, two-dimensional radial-flow-model used to simulate the Agua Caliente Spring was about 800 ft (244 m) thick and extends about 3,280 ft (1,000 m) radially beyond the spring. The grid contains 2,841 elements and telescopes radially, starting at a width of about 2.5 ft (0.75 m) for the first column of grid elements, which represents the spring orifice, increasing by 5 ft (1.5 m) increments for the next 18 columns of grid elements, and then widening the grid elements by a factor of about 1.3 to a maximum of 660 ft (200 m) at the furthest extent of the flow model (fig. 29). Vertically, the model was divided into 73 variably spaced rows. The vertical grid spacing starts at about 3.3 ft (1 m) increments for the top 9 rows of grid elements, reduces to about 1.6 ft (0.5 m) increments for the next 2 rows (coinciding with the perching layer), increases to about 3.3 ft (1 m) increments for the next 4 rows, increases to about 13 ft (4 m) increments for the next 57 rows, and finally is reduced to about a 6.6 ft (2 m) increment for the bottom

row (fig. 29). The bottom boundary of the flow domain is an assumed no-flow interface between the valley-fill deposits and the basement complex. The water table is 216 ft (66 m) bls. The lateral boundaries of the model domain are fixed no-flow boundaries with a constant saturation of 1.00 starting at 216 ft (66 m) bls to a depth of about 800 ft (244 m) and with a temperature set at 79.2 °F (26.2 °C), chosen to closely match the water temperature in well 4S/4E-14Q1 (table 7). This boundary is far enough away from the spring so that the boundary condition does not limit the simulated heat loss from the spring chimney. The upper boundary condition is set to one atmosphere at 79.2 °F (26.2 °C), which is approximately the air temperature measured in the well casing of well 4S/4E-11Q1 at about 20 ft (6 m) bls (fig. 30). The upper boundary condition allows for heat exchange with the upper soil layer. The interface between land surface and the atmosphere is a no-flow boundary to water; however, heat and air are allowed to flow through the interface. A specified-flux boundary condition was used in column 1, at the bottom boundary of the model (fig. 29), to simulate the Agua Caliente Spring. The rate and temperature of the specified-flux boundary was adjusted during the model calibration. The temperature of the regional groundwater was set to 79.2°F (26.2°C) on the basis of water temperature data collected from well 4S/4W-14Q1 (table 7). The radial-flow model was run until steady-state conditions were reached (no changes in flow or heat within the model domain).

The three-dimensional flow model of the Agua Caliente Spring is 216 ft (66 m) on a side and 216 ft (66 m) deep (fig. 31). The model extends from land surface to the top of the regional water table, 216 ft (66 m) bls (fig. 31C). The model layers are horizontal with zero slope, and the model domain contains 49,393 grid elements. The model discretization in the X and Y directions is variable, with the smallest columns of 1.6 ft (0.5 m) representing the Agua Caliente Spring and a possible secondary orifice of the Agua Caliente Spring not contained by the steel collector tank, near the center of the grid (fig. 31B). The model has the same vertical discretization as the upper 216 ft of the radial-flow model (figs. 29B and 31C). The upper boundary condition is set to one atmosphere at 79.2°F (26.2°C); the same as in the radial-flow model). The lateral boundaries of the model domain are fixed no-flow boundaries that are specified to be in gravity potential equilibrium from the water table (set at zero potential or saturation) to the land surface. The bottom boundary of the flow domain is a zero potential head boundary, representing the water table.

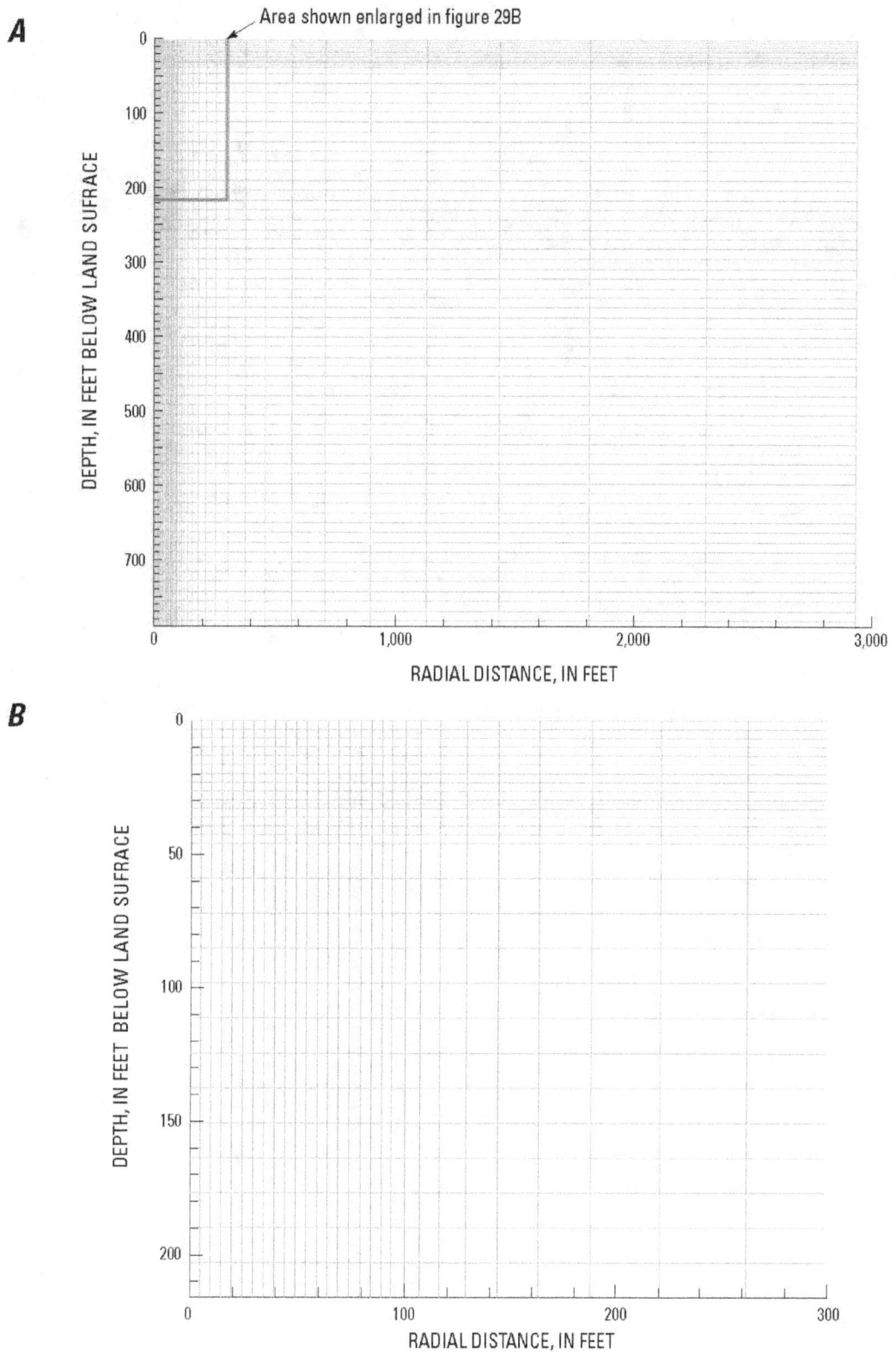

Figure 29. Grid discretization for (*A*) entire model domain and (*B*) the upper 197 feet (60 meters) of the model domain for the radial flow model of the Agua Caliente Spring, California.

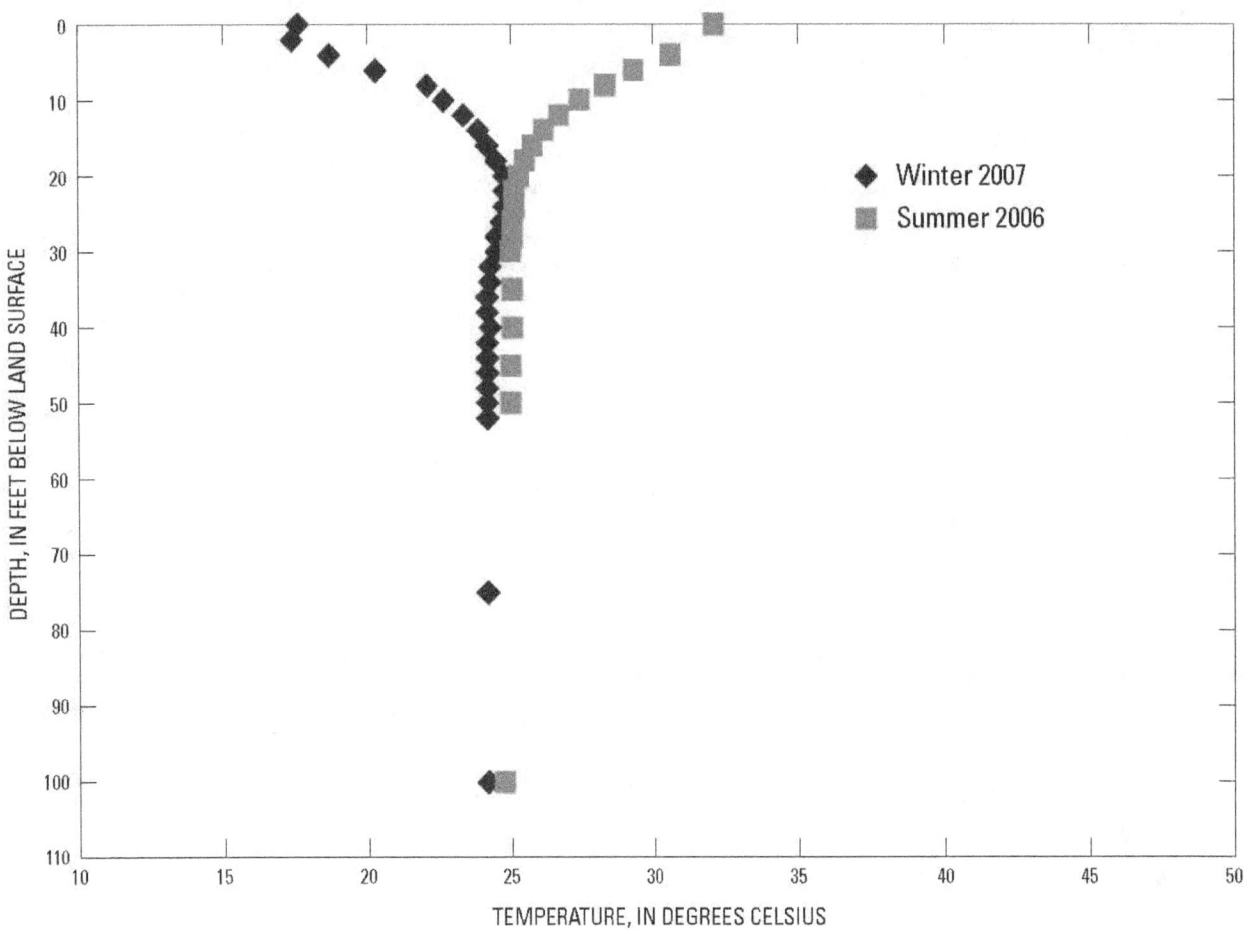

Figure 30.   Measured downhole temperature profile for well 04S/04E-11Q01 during summer 2006 and winter 2007.

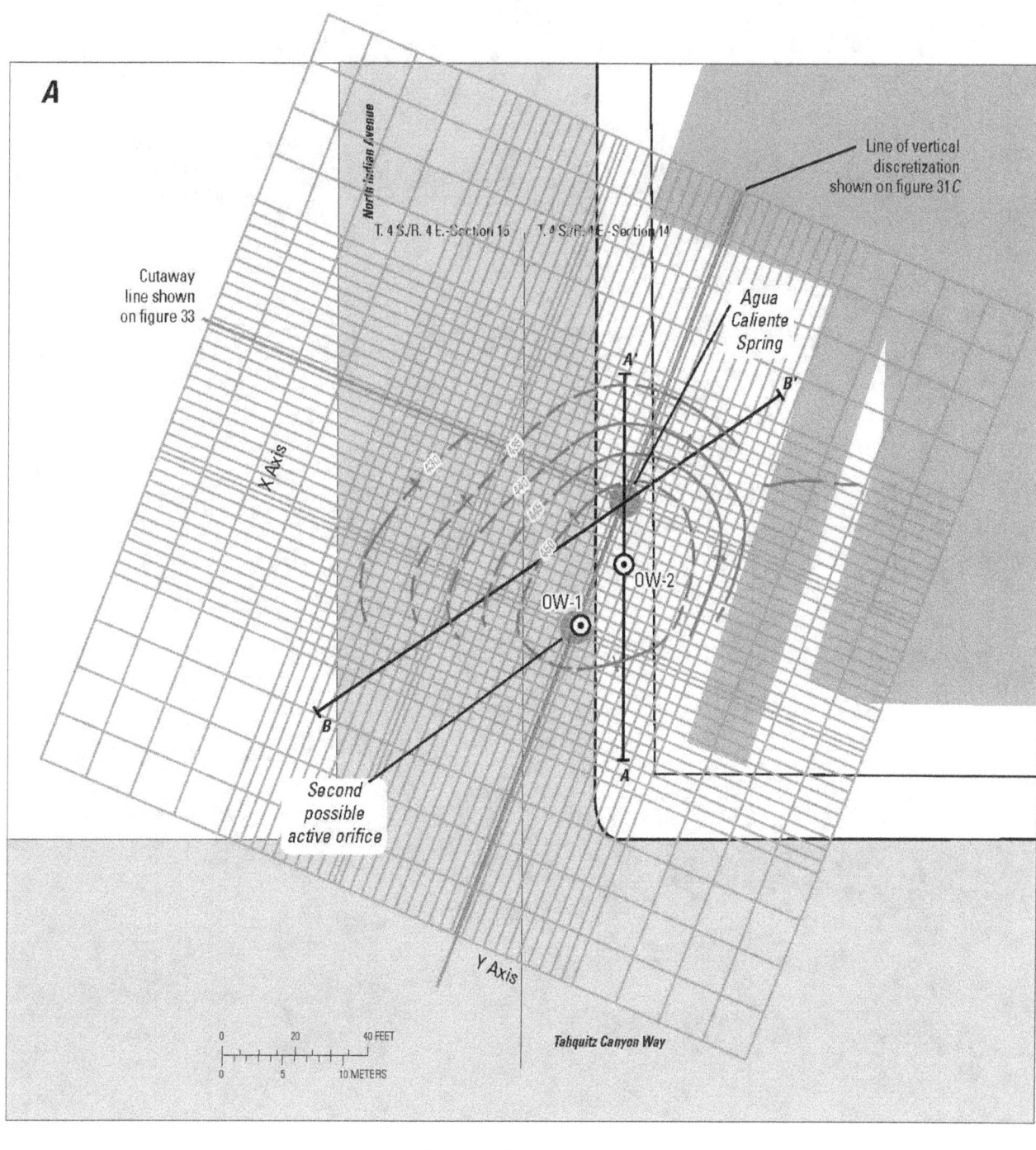

EXPLANATION

Line of equal water level
from figure 5B (Jan, 1959).
Dashed where approximate

Figure 31. (A) location, (B) discretization along the X and Y directions, and (C) discretization along the Z direction for the three-dimensional model of the area surrounding the Agua Caliente Spring, California.

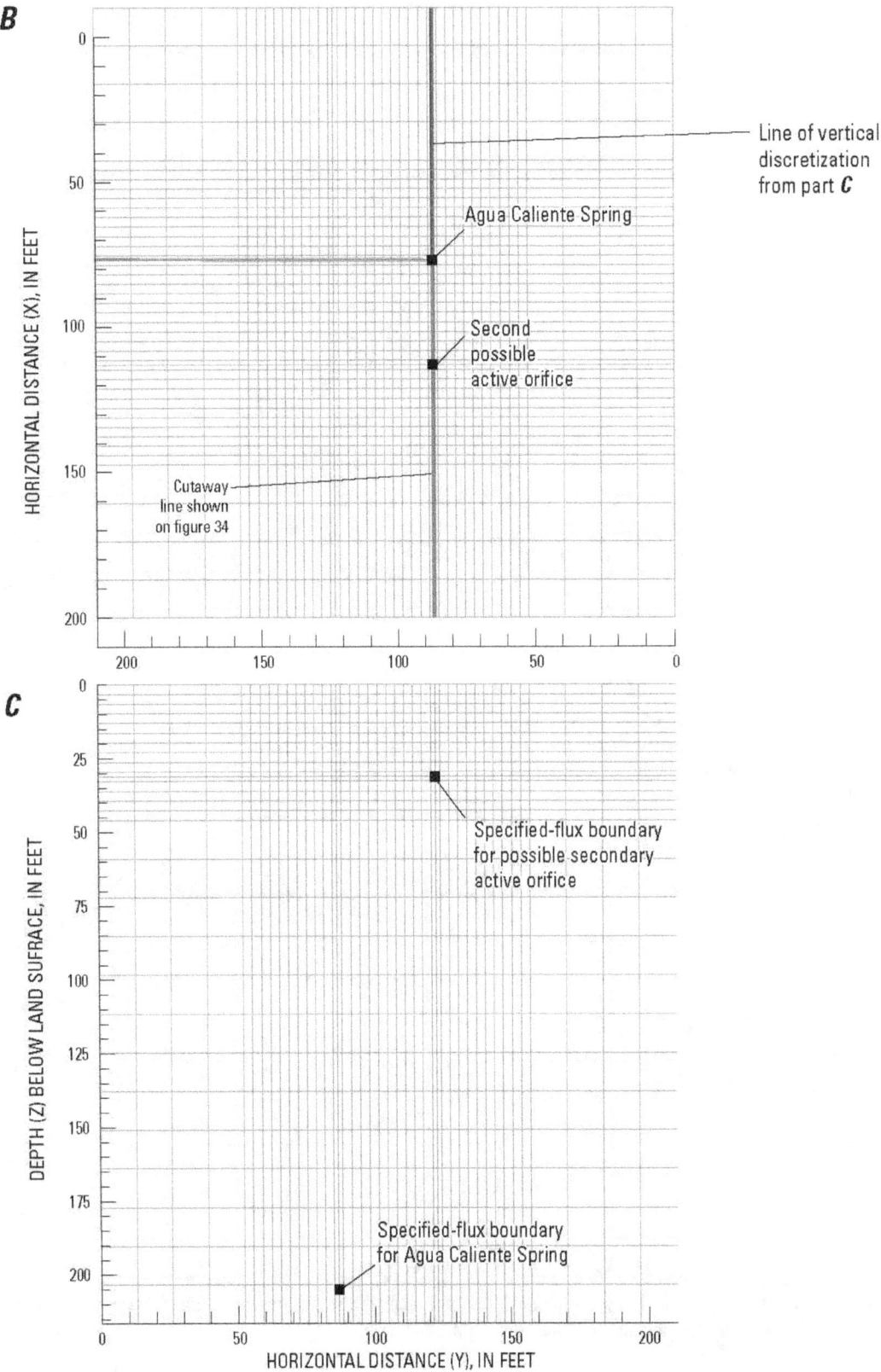

Figure 31.—Continued

Model Properties

The geologic materials in the study area were generalized into five main units: (1) spring orifice deposits, (2) chimney deposits (materials immediately adjacent to the orifice deposits), (3) peripheral materials, (4) perching layer, and (5) valley-fill deposits. Dutcher and Bader (1963) reported that the orifice deposits consist of silty-fine sand with a laboratory measured hydraulic conductivity of about 10.7 ft/d. Dutcher and Bader (1963) reported that the peripheral material consists of very fine sand, silt, and clay deposits that are present within a radius of 50 ft of the spring orifice. Dutcher and Bader (1963) postulated that the spring water washed the fine material (fine sand, silt, and clay) from the valley-fill deposits and then deposited this fine material in the area surrounding the spring orifice, resulting in the fine-grained character of the peripheral material. The laboratory measured hydraulic conductivity of the peripheral materials was about 0.05 ft/d (Dutcher and Bader, 1963). Inspection of available geologic logs indicates that there is a low-permeability layer of silt and clay at about 30 ft bls, referred to in this report as the perching layer. No laboratory measurements of hydraulic conductivity were available for the perching layer. The valley-fill deposits consist of sands and gravels with hydraulic conductivity values ranging from about 1 to 50 feet per day (ft/d; Reichard and Meadows, 1992). Thermal properties of the unsaturated zone materials were measured in the USGS soil physics laboratory in Sacramento, California (table 11).

Hydraulic and thermal properties were assigned to each of the five geologic units on the basis of published, measured, or estimated values (table 11). Initial model runs indicated that the model results were most sensitive to changes in the simulated hydraulic conductivity of the spring orifice, spring chimney, and perching layer. Therefore, model calibration involved modifying the initial estimates of hydraulic conductivity for the spring orifice, spring chimney, and perching layer using a trial-and-error process, while keeping the values of the other parameters the same as the initial estimates. The spring orifice was simulated by the vertical hydraulic conductivity in the model cells representing the spring, and the spring chimney was simulated by the horizontal hydraulic conductivity in the spring cells. The calibration process involved adjusting the initial estimates of these parameters to improve the match between the simulated and measured rate of Agua Caliente Spring discharge, temperature of the Aqua Caliente Spring discharge, and lateral extent of perched water. The measured calibration targets for the rate of spring discharge, the temperature of spring discharge, lateral extent of perched water, and temperature of the regional aquifer were 25 gal/min (the Agua Caliente discharge rate recorded by Dutcher and Bader [1963; table 4] in 1959); 104°F (40.0°C; the mean water

temperature measured in wells OW-1 and OW-2, fig. 32); 50 ft (the maximum lateral extent of perched water measured by Dutcher and Bader, 1963); and 79.2°F (26.2°C), measured in well 4S/4W-14Q1 (table 7), respectively.

The simulated temperature in the area around Agua Caliente Spring was constrained by downhole temperature profiles collected during the summer of 2006 and the winter of 2007 in wells OW-1, OW-2, and 4S/4E-11Q1 (figs. 30 and 32) and one historical measurement collected at well 4S/4E-15J1 (table 7). Wells OW-1 and OW-2 are adjacent to the Agua Caliente Spring, well 4S/4E-11Q1 is approximately 0.7 mi northeast of the spring, and well 4S/4E-15J1 is approximately 800 ft southwest of the spring (fig. 1). Temperatures at OW-1 and OW-2 were recorded below the perched water table, with the exception of a few measurements that were collected above the perched water table, whereas temperatures in well -11Q1 were recorded in the air, above the regional water table. One temperature was measured from well -15J1 while it was pumping in 1978. Temperatures measured at OW-1 at 2 ft bls were 110.3°F (43.5°C) in the summer and 81.9°F (27.7°C) in the winter. During the summer, temperatures decreased with depth to approximately 10 ft bls, then remained stable to the bottom of the well at 37 ft; in contrast, winter temperatures increased with depth until approximately 28 ft bls and then remained constant to the bottom of the well. Temperatures measured at 4S/4E-11Q1 near the land surface were 89.8°F (32.1°C) in the summer and 63.3°F (17.4°C) in the winter. During the summer, temperatures measured in well 4S/4E-11Q1 decreased with depth to about 20 ft bls, whereas winter temperatures increased with depth to about 20 ft bls. Temperatures remained fairly constant below 20 ft bls to the maximum depth measured at 100 ft bls. The one temperature measured from the regional water table at well 4S/4E-15J1 was 79.2°F (26.2°C) in 1978 (table 7).

The hydraulic and thermal properties specified or calibrated for the spring orifice, spring chimney, peripheral material, valley-fill deposits, and perching layer in the radial-flow model were used in the three-dimensional model without modification. A second spring orifice of the Agua Caliente Spring, not contained by the steel collector tank, was simulated approximately 36 ft (11 m) away from the Agua Caliente Spring in the three-dimensional model (fig. 31). The second orifice was assumed to branch off from the Agua Caliente Spring orifice above the perching layer (fig. 64). The second spring orifice had the identical hydraulic properties as the Agua Caliente Spring orifice above the perching layer, except the upper model element was assigned the hydraulic properties of the valley-fill deposits. This allowed the spring discharge from the second orifice to flow into the perched aquifer.

Table 11. Hydraulic and thermal properties used in the radial and three-dimensional model simulations of the Agua Caliente Spring, California.

[Btu, British thermal units; °F, degrees Farenheit; ft, feet; h, hours, lb, pounds; ft², square feet; ft/d, feet per day; m, fitting function for the Van Genuchten equation, unitless; α, air entry pressure, per foot]

Material	Depth interval (ft)	Radial interval (ft)	Porosity	Thermal conductivity (Btu·ft·(h·ft²·°F)⁻¹)	Specific heat (Btu·(lb·°F)⁻¹)	Saturated hydraulic conductivity		Van Genuchten parameters	
						Initial (ft/d)	Calibrated (ft/d)	(m)	(−1/α)(ft)
Spring orifice	0–800	0–5	0.48	0.9	0.24	10.7	220	0.412	1.5
Chimney deposits	0–800	0–5	0.48	0.9	0.24	0.05	0.02	0.412	1.5
Peripheral material	0–26 30–800	5–50	0.47	0.9	0.40	0.05	1.1	0.435	16.0
Perching layer	26–30	5–3000	0.35	0.9	0.24	0.2	0.2	0.412	1.5
Valley-fill deposits	0–26 30–800	50–3000	0.35	0.9	0.24	5.6	5.6	0.412	1.5

A

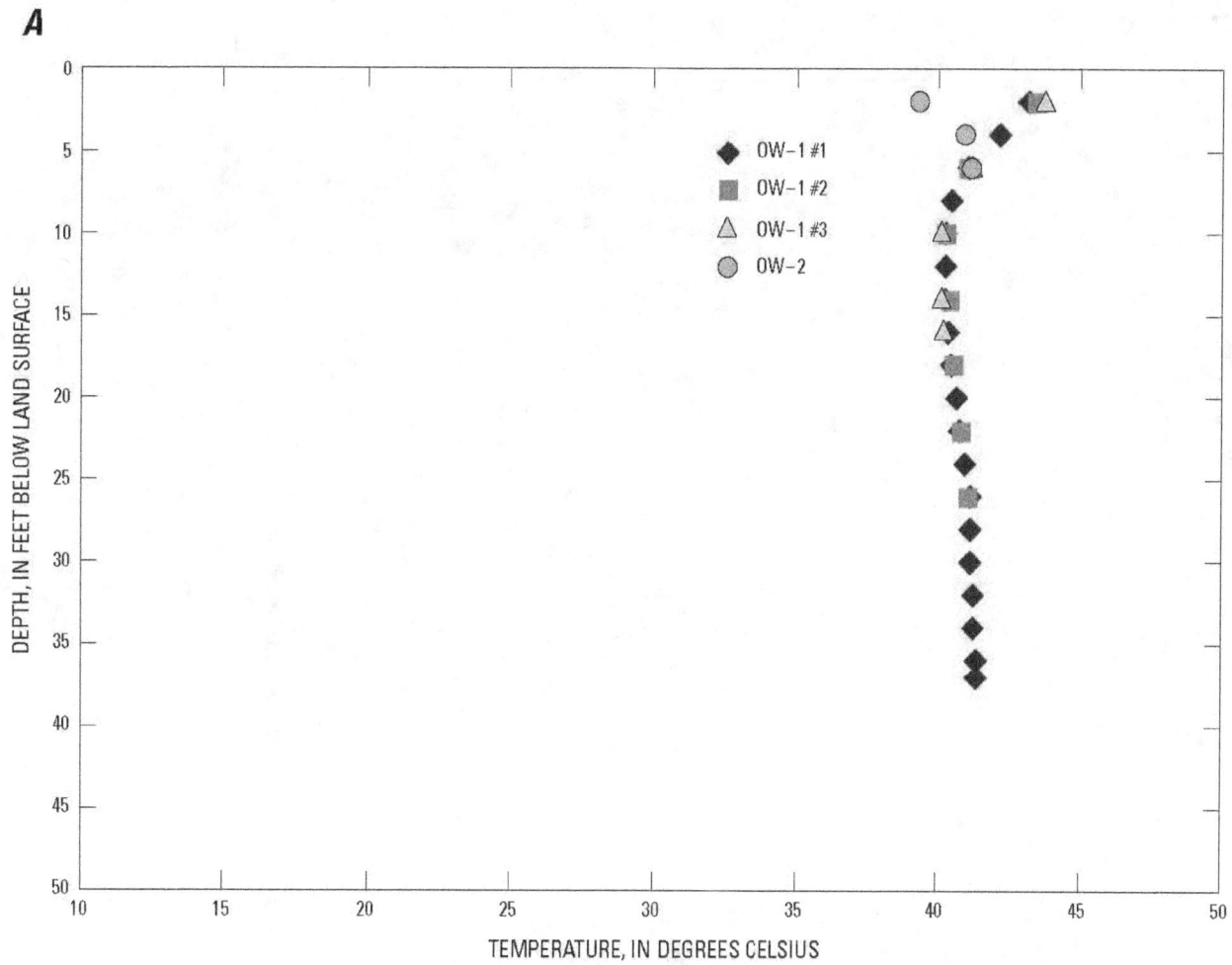

Figure 32. Measured downhole temperature profile in wells OW-1 and OW-2 during (*A*) summer 2006 and (*B*) winter 2007.

B

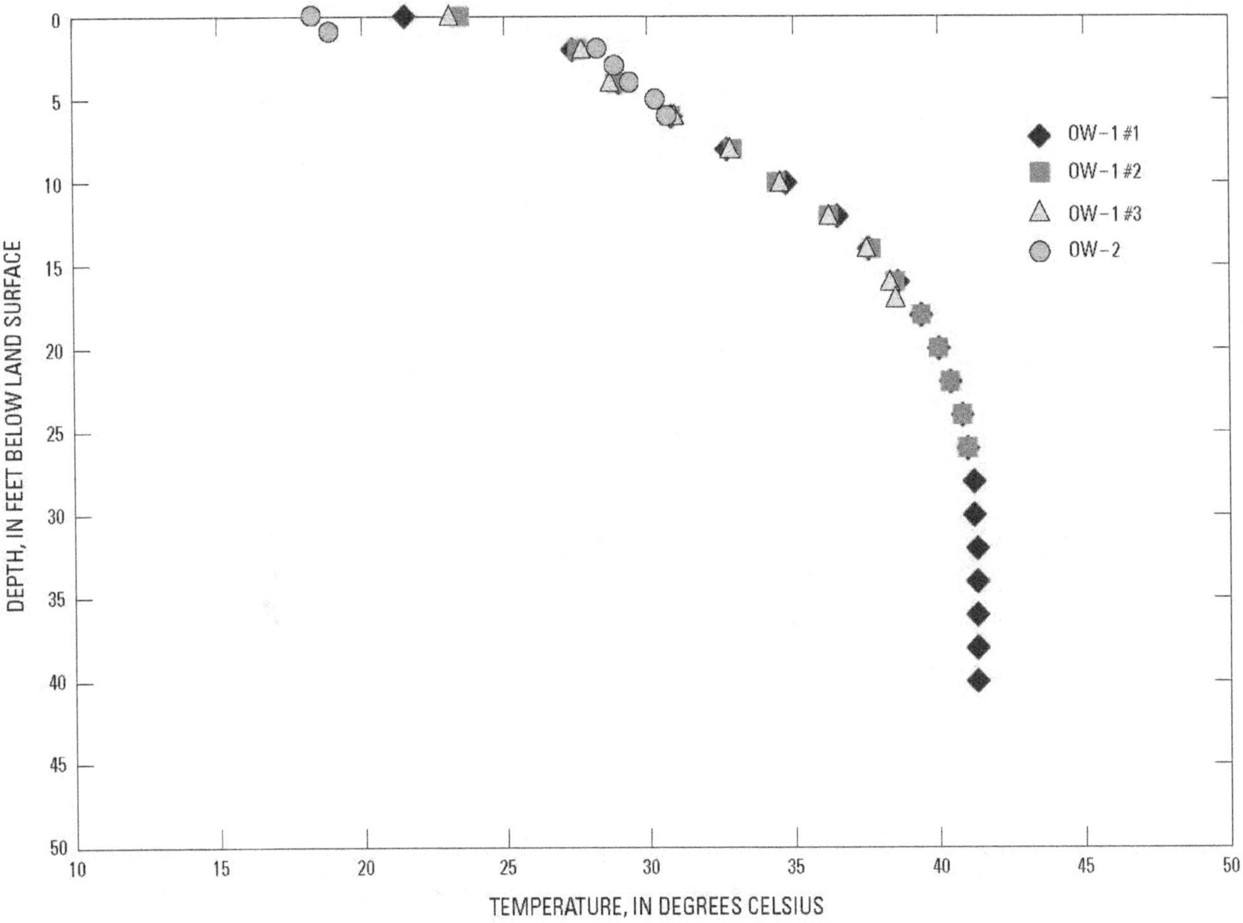

Figure 32.—Continued

Model Results

Testing the Conceptual Model of the Agua Caliente Spring Chimney

As stated previously, Dutcher and Bader (1963) postulated that the Agua Caliente Spring enters the valley-fill deposits from fractures in the underlying basement complex (fig. 10). Because the spring orifice in the upper 12 ft did not show any evidence of cementation, they believed that the spring water rises through a washed-sand conduit of its own making and that the lateral movement of the spring is restricted by low-permeability, sandy-silty clay that was deposited or reworked from the valley-fill deposits. The low-permeability reworked deposits immediately adjacent to the spring orifice are referred to as the spring chimney by Dutcher and Bader (1963). On the basis of available well logs and gravity data collected for this study, the spring rises through more than 800 ft of valley-fill deposits. Seismic profiles collected for this study (fig. 17) indicate that the valley-fill deposits become indurated at about 200 to 500 ft bls near the Agua Caliente Spring. Therefore, the spring may only rise 200 ft through the unconsolidated valley-fill deposits, if faulting along the buried ridge of basement complex and indurated valley-fill deposits provides a pathway for the deep thermal water to rise into the unconsolidated valley fill deposits. For the purposes of the modeling completed for this project, however, it was assumed that the entire thickness of the valley-fill deposits was unconsolidated.

The radial-flow model was used to test the conceptual model of the spring chimney. The model was used to test what hydraulic conductivity values would be required for the spring orifice and chimney in order for the model to simulate the measured discharge rate and temperature of the spring at the current (2008) land surface. As previously stated, a specified-flux boundary condition was used in column 1, at the bottom boundary of the model (about 800 ft bls), to simulate the Agua Caliente Spring (fig. 29). For the purposes of the modeling exercise, the measured rate and temperature of the

Agua Caliente Spring at land surface were chosen to be 25 gal/min and 104°F (40°C), respectively. Initially, these values were used as input to the bottom boundary of the model; however, as described later in the "Source of Perched Water" section, a second spring orifice discharging 5 gal/min was required to maintain the observed distribution and temperature of water in the perched aquifer near the Agua Caliente Spring. The second orifice was assumed to branch off from the Agua Caliente Spring orifice above the perching layer; therefore, the total rate of discharge simulated at the bottom boundary of the model had to be at least 30 gal/min. The rate and temperature of the specified-flux boundary was adjusted slightly during the model calibration, until the simulated rate and temperature matched the measured values.

The peripheral material, valley-fill deposits, and perching layer were assumed to be isotropic, having hydraulic conductivity values of about 1.0 ft/d, 6.0 ft/d, and 0.2 ft/d, respectively. As stated previously, the hydraulic conductivity of the peripheral material and valley-fill deposits were not modified during the model calibration. The simulated hydraulic conductivity values for the peripheral material and valley-fill deposits probably are poor estimates of actual values because they are simulated as being isotropic, and there are few data to constrain the simulated values. Assuming isotropy for the peripheral material and valley-fill deposits does not significantly affect the model results because these parameters are relatively insensitive to reasonable changes in their values. The hydraulic conductivity of the perching layer was initially assumed to be one-tenth of the hydraulic conductivity of the peripheral material. The initial estimate was increased by about two times to simulate the lateral extent of the observed perched aquifer. The calibrated hydraulic conductivity of the perching layer is representative of the vertical hydraulic conductivity of this unit because it restricts the downward flow of the discharge from a second spring orifice. The low hydraulic conductivity of the perching layer (0.2 ft/d) allowed the perched aquifer to occur, but also allowed enough infiltration to keep the perched aquifer from becoming laterally extensive.

The model was most sensitive to the hydraulic conductivity of the spring orifice (simulated by the vertical hydraulic conductivity of the spring cells) and the spring chimney (simulated by the horizontal hydraulic conductivity of the spring cells). The calibrated vertical hydraulic conductivity of the spring cells (spring orifice) was about 200 ft/d, and the horizontal conductivity of the spring cells (spring chimney) was about 0.00002 ft/d. The calibrated hydraulic conductivity of the spring orifice was about 20 times greater than the laboratory measurement of spring orifice deposits (about 10.7 ft/d; Dutcher and Bader, 1963) collected during the excavation of the spring in 1958. Higher values could be simulated and produce reasonable results, but lower values resulted in simulated heads that were too high at land surface. Higher values of vertical hydraulic conductivity probably are not reasonable on the basis of the lithologic description of the spring orifice deposits (silty fine sands; Dutcher and Bader, 1963). The calibrated hydraulic conductivity of the spring chimney was about three orders of magnitude lower than the laboratory measurement of peripheral material (about 0.05 ft/d; Dutcher and Bader, 1963) collected during the excavation. The simulated hydraulic conductivity of the spring chimney could be lower than the calibrated value, but higher values caused significant leakage from the spring to surrounding peripheral material and valley-fill deposits, and could not support the measured spring discharge. In addition, leakage from the spring resulted in heating of the regional aquifer near the spring. The quantity of spring leakage was constrained by the temperature of the regional aquifer, measured in well 4S/4E-15J1 [79.2°F (26.2°C), table 7]. The calibrated hydraulic conductivity of the spring chimney is about two orders of magnitude lower than the reported hydraulic conductivity of layered clay deposits (Bear, 1972), suggesting that the spring chimney may be cemented at depth.

The simulated rate and temperature of the specified flux for the calibrated model were 30.3 gal/min and 104.5°F (40.3°C), respectively. The simulated discharge rate and temperature of the Agua Caliente Spring at land surface were 30.1 gal/min and 104°F (40.0°C), respectively. The model simulates that the specified spring discharge at the bottom of the model loses about 0.2 gal/min and about 0.5°F (0.3°C) as the spring flows towards land surface. The low simulated loss

in spring discharge supports the water-quality data presented in the "Water-Quality Results" section that indicate an absence of leakage or mixing from the regional aquifer into the thermal Agua Caliente Spring.

The model results support the basic conceptual model presented by Dutcher and Bader (1963) for the development of the Agua Caliente Spring; however, the low calibrated hydraulic conductivity value for the spring chimney suggests that the chimney may be cemented at depth. Results of a geochemical model described in the "Saturation Controls on Chemical Composition" section indicate that the water at Agua Caliente Spring is at saturation with respect to both calcite and chalcedony, which would provide a possible mechanism for cementation of the spring chimney. A monitoring site in the regional aquifer near the Agua Caliente Spring, suitable for collecting depth-dependent aquifer property, water-level, chemical, and temperature data from the regional aquifer, would help constrain the model and improve the reliability of the model results.

Water-Level Declines in the Regional Aquifer

There is concern that water-level declines in the regional aquifer in the area of the Agua Caliente Spring could influence the temperature of the spring discharge. To evaluate this concern, the radial-flow model was used to simulate steady-state conditions with the regional water table at 216 ft (66 m) bls (the depth to the regional water table in 2007; fig. 33A) and at 316 ft (96 m) bls (100-ft decline; fig. 33B). In both simulations, about 4.5 gal/min of thermal water was allowed to leak from the steel collector tank at about 3.3 ft (1.0 m) bls to create a perched water table. The model simulates convection cells around the spring chimney in the saturated zone (fig. 33A and B). The water in the saturated zone is warmed by heat loss from the spring, causing it to rise. As the water rises, colder water moves to replace the water that has been heated by the spring, developing the convection cell. The simulated decline in the water table had no effect on the spring discharge and resulted in a slight increase in the temperature of the spring discharge (less than 0.1°C). The temperature increase is caused by a thicker unsaturated zone (fig. 33B) reducing the convective heat loss from the spring discharge.

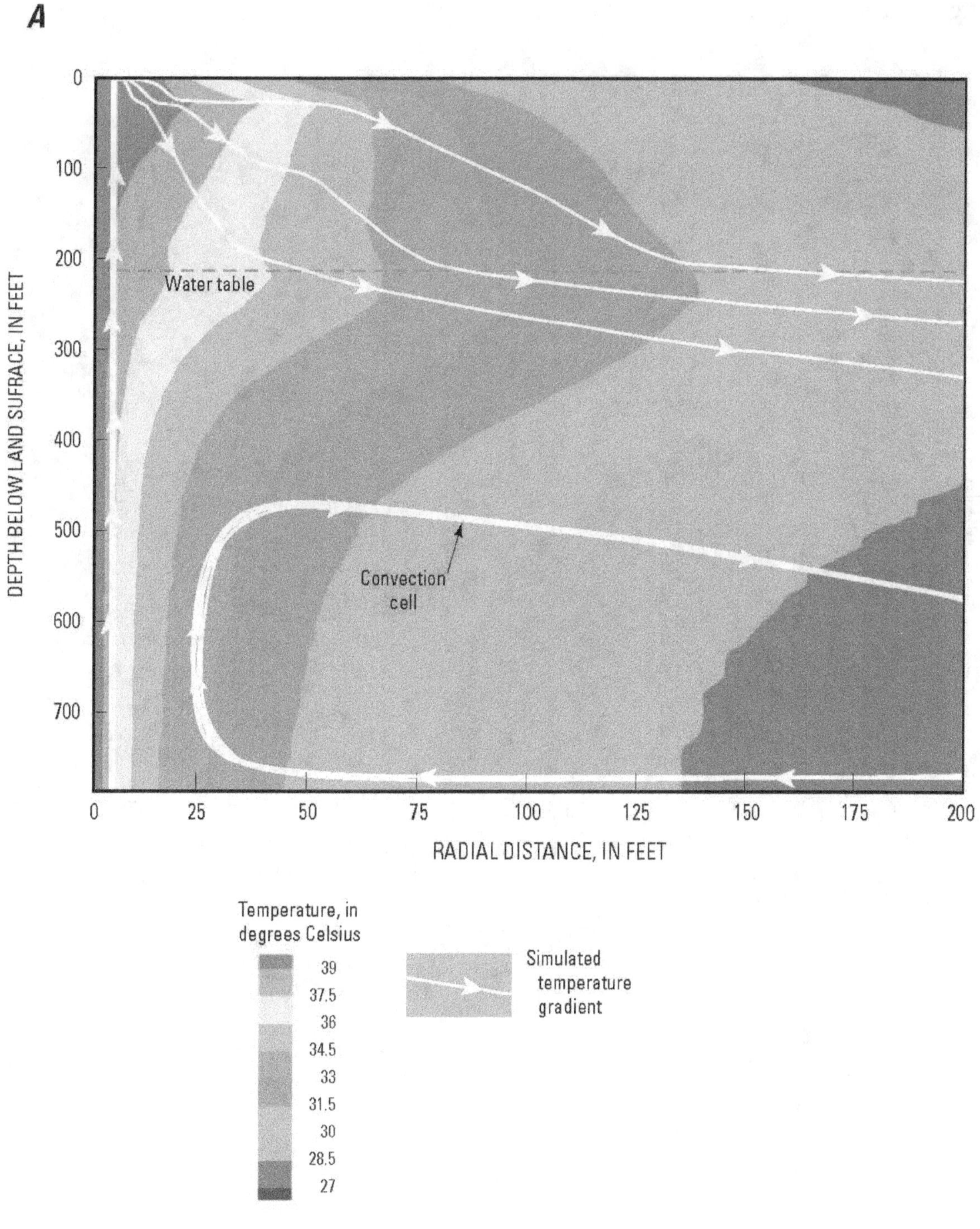

A

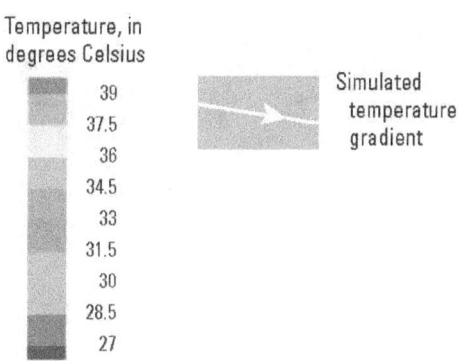

Figure 33.   Simulated steady-state conditions around the Agua Caliente Spring, California, (*A*) with the regional water table at 2007 conditions and (*B*) with the regional water table 100 feet lower than 2007 conditions.

B

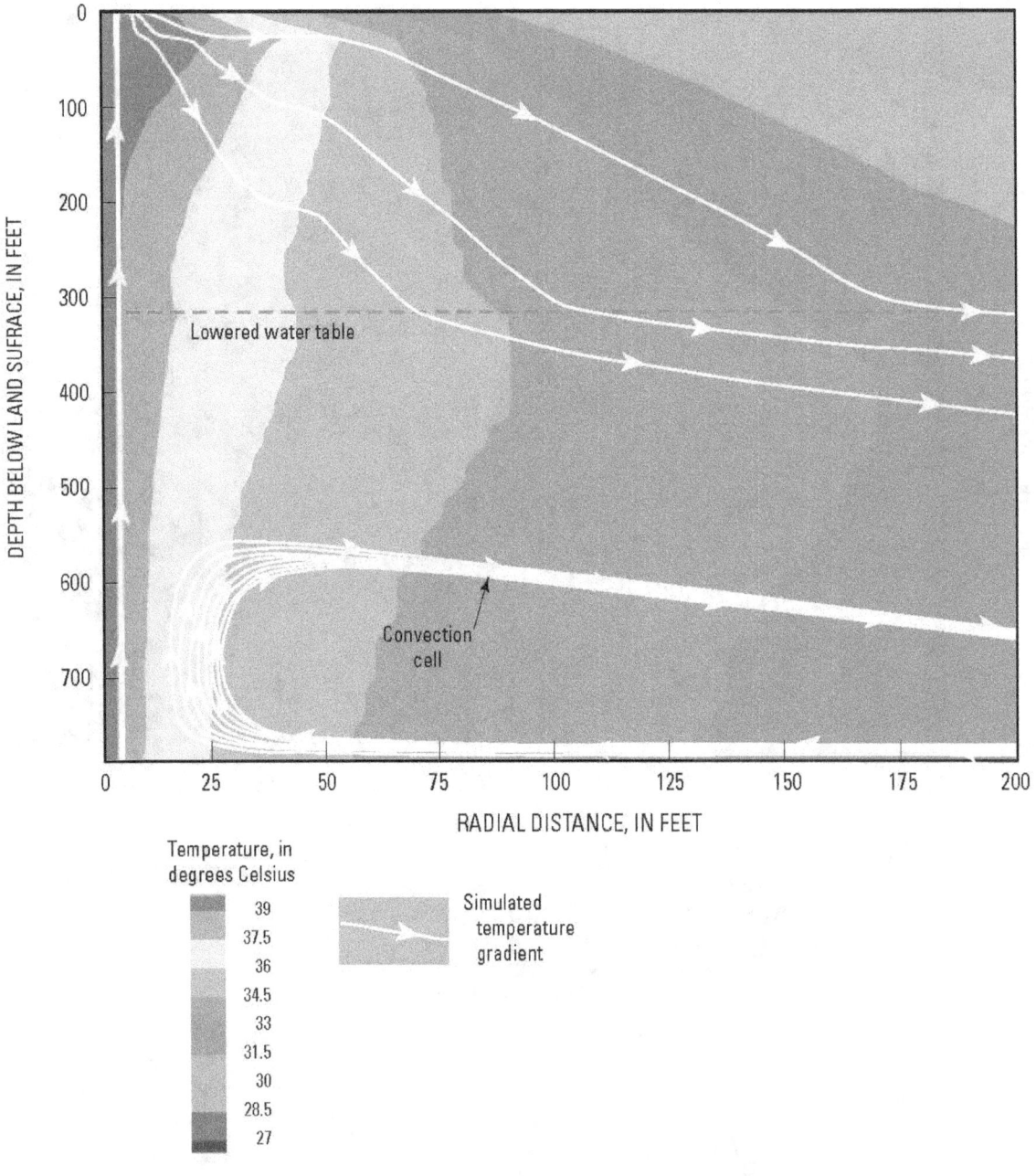

Figure 33.—Continued

Source of Perched Water

Dutcher and Bader (1963) documented the existence of a perched water table in the vicinity of the Agua Caliente Spring in 1951. A steel collector tank, open at the bottom, was installed above the spring orifice in 1959 to capture the spring discharge (fig. 3). However, even after the emplacement of the steel collector tank, perched water was present in the vicinity of the spring (fig. 6B). Data collected from monitoring sites OW-1 and OW-2 for this study document that the perched water was present in 2007 (table 1). The calibrated radial-flow model was used to test three scenarios that could be responsible for developing the perched aquifer near the Agua Caliente Spring: (1) leakage of thermal water from the steel collector tank at 3.3 ft (1.0 m) bls, (2) leakage of thermal water from the bottom of the steel collector tank at 11.5 ft (3.5 m) bls, and (3) leakage of thermal water from the spring chimney at 23 ft (7.0 m) bls. In all scenarios, the perched aquifer develops, but none of the scenarios produced a water-level mound in the perched aquifer near the spring that was reported by Dutcher and Bader (1963). These model simulations indicate that leakage from the steel collector tank cannot explain the altitude and temperature of perched water observed at test well OW-1.

The three-dimensional model was used to determine if a second spring orifice of the Agua Caliente Spring, not contained by the steel collector tank, could produce the observed shape and temperature of the perched water table surrounding the Agua Caliente Spring. The radial-flow model could not be used for this analysis because the simulation of the Agua Caliente Spring and a second spring orifice is not symmetric. The hydraulic and thermal properties specified or calibrated for the spring orifice, peripheral material, valley–fill deposits, and perching layer in the radial-flow model were used in the three-dimensional model without modification. A second spring orifice was simulated about 36 ft (11 m) away from the Agua Caliente Spring, in the approximate location of test well OW-1 (fig. 31). The second orifice was assumed to branch off from the Agua Caliente Spring orifice above the perching layer (fig. 64). A specified-flux boundary condition was used at the model elements representing the second spring orifice, in the model layer directly above the perching layer (fig. 31). The second spring orifice had the identical hydraulic properties as the Agua Caliente Spring orifice above the perching layer, except the upper layer of model elements representing the spring orifice were assigned the hydraulic properties of the valley-fill deposits. This allowed the spring discharge from the second orifice to flow into the perched aquifer. The rate of the specified-flux boundary was adjusted until the simulated extent of the perched aquifer was similar to the measured extent. The calibrated rate of the specified flux representing the second spring orifice was approximately 4.7 gal/min.

The simulated second orifice resulted in a water-level mound that spreads laterally away from the second spring orifice and vertically downward through the perching layer (fig. 34). Note that the perching layer (saturation above 0.95) extends beyond the chimney of the Agua Caliente Spring (fig. 34). The low flow rate from the second spring orifice maintains 104°F (40°C) at the base of the perching layer and allows the colder soil (asphalt) temperature in the winter to reduce the measured nearby well water temperature by over 10°C, as observed in the temperature data collected at test well OW-1 (fig. 32). These model results demonstrate that the thermal water in the perched water table can be explained by flow from a second spring orifice.

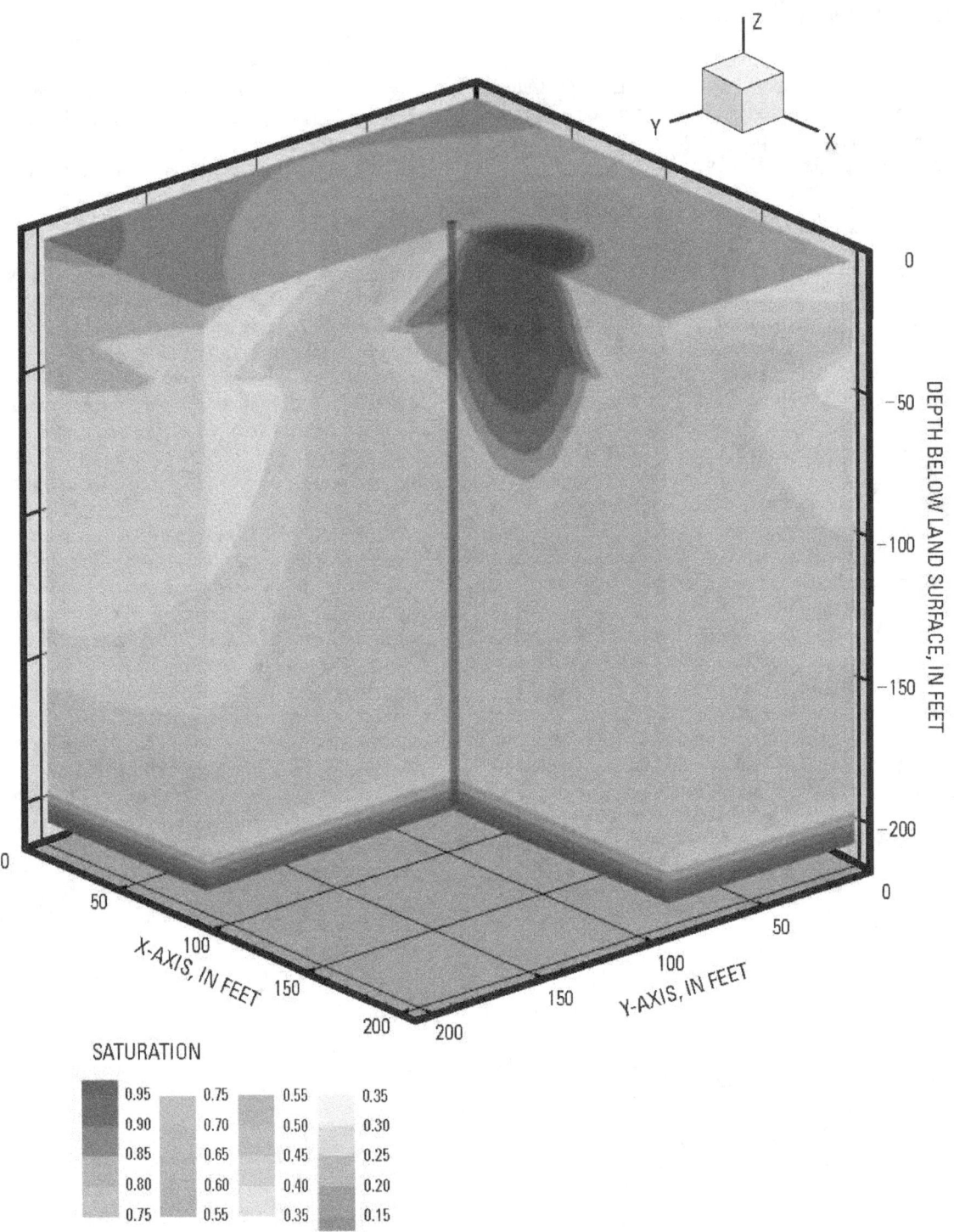

SATURATION

0.95	0.75	0.55	0.35
0.90	0.70	0.50	0.30
0.85	0.65	0.45	0.25
0.80	0.60	0.40	0.20
0.75	0.55	0.35	0.15

Figure 34. Simulated saturation in the area surrounding the Agua Caliente Spring, California, assuming a second spring orifice.

Summary and Conclusions

Agua Caliente Spring, in downtown Palm Springs, California, has been used for recreation and medicinal therapy for hundreds of years and is currently (2008) the source of hot water for the Spa Resort owned by the Agua Caliente Band of the Cahuilla Indians. The spring is located about 1,500 ft east of the eastern front of the San Jacinto Mountains on the southeast-sloping alluvial plain of the Coachella Valley. Geologic logs of water wells located between the spring and the mountain front indicate that the valley-fill deposits are at least 600 ft thick and the regional table is in excess of 200 ft bls in the vicinity of the Agua Caliente Spring. Data collected from shallow wells (less than 50 ft deep) indicate that a perched water table is present near the spring.

A gravity survey was conducted to define the thickness of the valley-fill deposits, or depth to the basement complex, beneath the Agua Caliente Spring area and to delineate geologic structures associated with the spring. Gravity data were collected at 252 gravity stations along traverses across Palm Canyon and the area of the Agua Caliente Spring. These data were combined with data previously collected from about 70 stations to develop a gravity field for the area. The gravity data indicate that the thickness of the valley-fill deposits is irregular along the western margin of the Coachella Valley, with a shallow buried ridge that strikes east-northeast as much as 10,000 ft away from the mountain front that appears to be a subsurface continuation of the steep ridge to the north of Tahquitz Canyon. The Agua Caliente Spring is located on the southeast flank of this buried basement ridge, where the gravity data indicate the valley-fill deposits are about 830 ft thick. The gravity field indicates that there is a north-south density boundary, along the eastern edge of the San Jacinto Mountains, which probably is caused by the continuation of the Palm Canyon fault. The Agua Caliente Spring is located within the inferred trace of the Palm Canyon fault, where the density boundaries suggest that the fault steps to the west along the buried basement ridge. Faulting of the basement complex along the buried ridge would allow deep thermal water to quickly rise into the overlying valley-fill deposits from an underlying reservoir of deep geothermal water, which is the probable source of the Agua Caliente Spring.

Shallow-depth seismic refraction and reflection surveys were conducted along three lines near the Agua Caliente Spring to help image the stratigraphy and geologic structures associated with the spring. Relatively high cultural noise levels in downtown Palm Springs did not allow velocity images to be developed below about 150 ft bls. Consistent with observations from nearby wells, interpretation of the seismic velocity images suggest that a perched groundwater table occurs in the upper 30 ft of sediments, north and east of the intersection of Indian Canyon Drive and Tahquitz Canyon Way. The seismic reflection data indicate that the depth to the basement complex is about 830 ft bls directly beneath the Agua Caliente Spring, and that the depth to the basement complex decreases from south to north indicating the presence of a buried basement ridge to the north of the Agua Caliente Spring, similar to the gravity data. The migrated seismic reflection images indicate the presence of a density contrast above the seismic interpreted depth to basement complex, which is interpreted as the contact between overlying unconsolidated valley-fill deposits and underlying indurated valley-fill deposits. The seismic interpreted contact between the unconsolidated valley-fill deposits and the indurated valley-fill deposits is about 500 ft bls directly beneath Agua Caliente Spring and rises to about 200 ft bls less than 500 ft east and north of the spring. The migrated seismic reflection images show disruptions in the layering and changes in the character of reflectors in the strata beneath the spring, suggesting the presence of two faults, which probably are related to the north-south trending Palm Canyon fault. Faults were not apparent on the seismic reflection image collected to the north of the Agua Caliente Spring, suggesting that the Palm Canyon fault steps to the west near the spring, as was inferred from the gravity data. This approach also found that faulting of the basement complex along the buried ridge of basement complex and indurated valley-fill deposits could provide a pathway for the deep thermal water to rise to about 200 ft bls near the Agua Caliente Spring from an underlying reservoir of deep geothermal water, which is the probable source of the Agua Caliente Spring.

InSAR was used in this study to help identify ground-surface deformation and locate structures, such as faults, that may affect groundwater movement. Analyses of 18 interferograms representing time periods ranging from 35 to 595 days between October 26, 2003, and September 25, 2005, indicate that little deformation (less than 0.6 in.) has occurred in the study area for the time periods represented by the interferograms. While this area was fairly stable for the time periods represented by the interferograms, a linear deformation boundary about 1 mi west of the Palm Springs International Airport is shown in three interferograms, indicating measureable subsidence. The deformation boundary trends northwest-southeast and is near the location of a fault inferred to be present from other studies. None of the interferograms provide information on the location of possible buried faults near the Agua Caliente Spring.

Historical measurements indicate that the Agua Caliente Spring discharge has varied between 5 and 60 gal/min over the past century. Possible explanations for the historical variation include differences in methods used to measure discharge, fluctuations in climate, and(or) changes of the spring discharge area over time—from an uncontrolled discharge into a pond in the early 1900s to a controlled discharge into a steel collector tank since 1958. Data collected for this study indicate that the discharge varied from a high of about 24 gal/min in the summer of 2005, following 2 years with above-normal precipitation, to a low of about 9 gal/min in the summer of 2006, following a year with below-normal precipitation. These observations suggest that the discharge of Aqua Caliente Spring is influenced by recent precipitation, although discharge data need to be collected over a period spanning multiple wet and dry cycles to establish the relationship with a high degree of confidence.

Available records indicate that the temperature of the Agua Caliente Spring has been relatively constant over the past century, ranging from a low of 37.8°C in 1917 to a high of 42.2°C in 1953. Water temperature measured at Agua Caliente Spring during this study was nearly constant, ranging from a low of 40.7°C to a high of 41.8°C between April 29, 2005, and September 12, 2006. Unlike the discharge of the Agua Caliente Spring, the temperature of the spring does not appear to be influenced by recent precipitation. The apparent influence of precipitation on discharge from the Agua Caliente Spring and lack of influence of precipitation on the temperature of the spring, can be explained by a piston-flow conceptual model for the spring, similar to the flow of water through a hose. Precipitation in the surrounding mountains recharges the underlying geothermal reservoir through fractures in the basement complex. The reservoir is large compared to the seasonal recharge contributed to the reservoir, and is surrounded by the low permeability basement complex. Water slowly moves through the reservoir and is heated as it flows toward an exit, such as a continuous fracture or fault. The fractures or faults are relatively small compared to the reservoir; therefore, as water recharges the reservoir, the discharge from the reservoir occurs almost simultaneously, similar to when water is added to a full hose. Because water movement through the reservoir is a long process, and the seasonal pulses are small compared to the total water stored in the reservoir, the variable discharge from the reservoir maintains a relatively constant temperature.

Available historical water-quality data and seasonal data collected during this study were used to define the source(s), and the age(s), of water discharged by the Agua Caliente Spring and to ascertain the seasonal and longer-term variability of chemical characteristics of the spring discharge. Sodium (Na) composes more than 95 percent of total cations in samples from the Agua Caliente Spring; whereas, it makes up only about 50 percent in samples from the regional aquifer. Another notable difference is that pH is nearly 10 (highly alkaline) in the Agua Caliente Spring but is only slightly higher than 7 (circum-neutral) in both regional groundwater on the valley floor and in the canyons. The great differences in sodium proportions and in pH are consistent with the absence of leakage or mixing from the regional aquifer into the thermal Agua Caliente Spring. Chemical composition changed very little, either seasonally or annually, from 2005 to 2006 in the Agua Caliente Spring, indicating an absence of response to changing discharge or precipitation that is consistent with a very old age for the water and absence of contribution from the regional aquifer. Comparison to historical data indicates water quality at the spring has not changed appreciably in the last 100 years.

Comparison of chemical concentrations between the Agua Caliente, Fenced, and Chino Warm Springs indicates differences are much greater for several trace elements than for major ions, hence a single common source for the geothermal water at the three sites is unlikely. Also, there are large differences in the trace element concentrations in the Agua Caliente and regional groundwater. Such large differences lend additional support to the inference made on the basis of major-ion water quality that no mixing occurs between the thermal water and the regional aquifer.

Temperature of the geothermal reservoir (geothermal source water) was estimated using two equations that are based on aqueous concentrations—the solubility of chalcedony and an empirical relationship between Na, K, and Ca. The empirical relationship yields a reservoir temperature between 61 and 71°C at Agua Caliente, Fenced, and Chino Warm Springs. This compares reasonably well with calculations based on aqueous equilibration with chalcedony, which are about 10 degrees lower for Fenced Spring, about 10 degrees higher for Chino Warm Spring, and about the same at Agua Caliente Spring. Both methods confirm a moderate temperature, far below the boiling point of water, for the geothermal source water at all three warm springs.

δD ranges from about −70 per mil in Fenced Spring to almost −80 per mil in Agua Caliente and Chino Warm Springs. The lighter (more negative) δD values in Chino Warm and Agua Caliente Springs are consistent with an older and(or) higher altitude source of recharge to these springs. The altitude of recharge was estimated using δD values of the spring discharge and published isotopic composition of precipitation at a monitoring station on Mt. San Jacinto. The altitude of recharge was estimated to be about 7,740 ft for Chino Canyon Creek, 7,260 ft for Chino Cold Spring, 7,750 ft for Chino Warm Spring, and 7,290 ft for Agua Caliente Spring. The calculated altitude of recharge is consistent with the recharge of precipitation in the San Jacinto Mountains being the source of water for the Agua Caliente Spring.

Dissolved N^2 and Ar concentrations, along with "excess air" concentrations, yield a calculated recharge temperature of about 15°C for Agua Caliente Spring, about 16°C for Fenced Spring, and about 19°C for Chino Warm Spring.

These calculated recharge temperatures are in agreement with recharge temperatures on the basis of noble gases of 14°C for Agua Caliente Spring, 15°C for Fenced Spring, and 19°C for Chino Warm Spring, and are consistent with the source of recharge to the springs being precipitation (snow and rain) in the higher altitudes of the San Jacinto Mountains.

Age of the water is arguably a spring's most important attribute when considering its susceptibility to climatic fluctuations and anthropogenic effects. Tritium is present at low concentrations in Chino Cold Spring and in a regional aquifer sample, indicating at least some contribution from water that is younger than 1950 (post-bomb). The extremely low value (0.1TU) of 3H at Agua Caliente Spring indicates almost no contribution of water younger than 1950, and this is consistent with the lack of mixing with groundwater in the regional aquifer that was indicated by the chemical characteristics of the spring and the surrounding aquifer. Carbon-14 activities for samples from the Agua Caliente, Chino Warm, and Fenced Springs range from 16 to 43 percent modern carbon. Calculated ^{14}C ages for three samples collected from the Agua Caliente Spring range from about 14,000 to 15,000 years before present, making its water the oldest of the three springs. This compares to somewhat younger ages of 10,000 years before present for a single sample from Chino Warm Spring and 7,000 to 8,300 years before present for three samples from Fenced Spring.

Numerical models of fluid and heat flow were developed for the Agua Caliente Spring to (1) test the validity of the conceptual model that the Agua Caliente Spring enters the more than 800-feet-deep valley-fill deposits from fractures in the underlying basement complex and rises through a washed-sand conduit and surrounding low permeability deposits (spring chimney) of its own making, (2) evaluate whether water-level declines in the regional aquifer will influence the temperature of discharging water, and (3) determine the source of thermal water in the perched aquifer. A radial-flow model was used to test the conceptual model and the effect of water-level declines. The observed spring discharge and temperature could be simulated if the vertical hydraulic conductivity of the spring orifice was about 200 ft/d and the horizontal hydraulic conductivity of the orifice (spring chimney) was about 0.00002 ft/d. The simulated vertical hydraulic conductivity of the spring orifice is within the range of values reported for sand; however, the low value simulated for the horizontal hydraulic conductivity of the orifice suggests that the spring chimney is cemented with depth. Chemical data collected for this study indicate that the water at Agua Caliente Spring is at saturation with respect to both calcite and chalcedony, which provide a possible mechanism for cementation of the spring chimney. A simulated decline of about 100 ft in the regional aquifer had no affect on the simulated discharge and resulted in a slight increase in the temperature of the spring discharge. Results from the radial-flow and three-dimensional models of the Agua Caliente Spring area demonstrate that the distribution and temperature of thermal water in the perched water table can be explained by flow from a second orifice of the Agua Caliente Spring not contained by the steel collector tank, but not by leakage from the collector tank.

The information gained from this study provides the Agua Caliente Band of the Cahuilla Indians a greater understanding of the hydrologic and chemical characteristics of the Agua Caliente Spring, enabling them to better preserve and manage this valuable resource. Continued monitoring of the flow and chemical characteristics of the spring would help to better define the relationship between climate and spring discharge and provide information on possible anthropogenic factors affecting the spring.

References Cited

Aeschbach-Hertig, W., Peeters, F., Beyerle, U., and Kipfer, R., 1999, Interpretation of dissolved atmospheric noble gases in natural waters: Water Resources Research, v. 35, p. 2779–2792.

Aeschbach-Hertig, W., Peeters, F., Beyerle, U., and Kipfer, R., 2000, Paleotemperature reconstruction from noble gases in groundwater taking into account equilibrium with entrapped air: Nature, v. 405, p. 1040–1044.

Aldrich, L.T., Wetherill, G.W., Tilton, G.R., and Davis, G.L., 1956, The half-life of ^{87}Rb: Physical Reviews, v. 104, p. 1045–1047.

Ball, J.W., and McCleskey, R.B., 2003, A new cation-exchange method for accurate field speciation of hexavalent chromium: Talanta, v. 61, p. 305–313.

Bear, J., 1972, Dynamics of Fluids in Porous Media, Dover Publications, 767 p.

Biehler, S., Langenheim, V.E., Sikora, R.F., Ponce, D.A., Chapman, R.H., Beyer, L.A., and Oliver, H.W., 2004, Bouguer Gravity Map of California—Santa Ana Sheet: California Geological Survey Map Series, scale 1:250,000.

Blakely, R.J., and Simpson, R.W., 1986, Approximating edges of source bodies from magnetic or gravity anomalies: Geophysics, v. 51, p. 1,494–1,498.

Brouwer, J., and Helbig, K., 1998, Shallow high-resolution reflection seismics. In: Helbig, K. and Treitel, S. (eds) Handbook of geophysical exploration, 19 Elsevier, Oxford, 391 p.

Brown, J.S., 1923, The Salton Sea region, California: U.S. Geological Survey Water-Supply Paper 497, 292 p.

Buick, R., 2006, RTK base station networks driving adoption of GPS +/- inch automated steering among crop growers: Trimble Navigation Limited, Trimble Agricultural Division, CO, white paper, 9 p.

Bullen, T.D., Krabbenhoft, D.P., and Kendall, C., 1996, Kinetic and mineralogic controls on the evolution of groundwater chemistry and $^{87}Sr/^{86}Sr$ in a sandy silicate aquifer, northern Wisconsin, USA: Geochemica Cosmochimica Acta, v. 60, p. 1807–1821.

California Department of Water Resources, 1979, Coachella Valley area well standards investigation: Los Angeles, California Department of Water Resources, Southern District, 40 p.

Catchings, R.D., Gandhok, G., Goldman, M.R., Okaya, D., Rymer, M.J., and Bawden, G.W., 2008, Near-surface location, geometry, velocities of the Santa Monica Fault zone, Los Angeles, California, Bull. Seis. Soc. Am., v. 98, p. 124-138 doi: 10.1785/0120020231

Catchings, R.D., Goldman, M.R., Gandhok, G., Horta, E., Rymer, M.J., Martin, Peter, and Christensen, Allen, 1999, Structure, velocities, and faulting relationships beneath San Gorgonio Pass, California: implications for water resources and earthquake hazards: U.S. Geological Survey Open-File Report 99–568, 53 p.

Catchings, R.D., Rymer, M.J., Goldman, M.R., and Gandhok, G., 2009, San Andreas Fault geometry at Desert Hot Springs, California and its effects on earthquake hazards and groundwater, Bull. Seis. Soc. Am., v. 99, p. 2190-2207 doi: 10.1785/0120080117

Coplen, T.B., 1994, Reporting of stable hydrogen, carbon, and oxygen isotopic abundances: Pure and Applied Chemistry, v. 66, p. 273–276.

Coplen, T.B., Wildman, J.D., and Chen, J., 1991, Improvements in the gaseous hydrogen-water equilibrium technique for hydrogen isotope analysis: Analytical Chemistry, v. 63, p. 910–912.

Corbaley, R.E. and Oquita, R., 1986, Geochemistry and geothermometry of the Desert Hot Springs geothermal resource area: Geothermal Resources Council, Transactions, v. 10, p. 107–112.

Craig, H., 1961, Isotope variations in meteoric water: Science, v. 133, p. 1702–1703.

Daly, C., Gibson, W.P., Doggett, M., Smith, J., and Taylor, G., 2004, Up-to-date monthly climate maps for the conterminous United States, Proceedings 14th AMS Conference on Applied Climatology, 84th AMS Annual Meeting Combined Preprints: American Meteorological Society, Seattle, Washington, January 13–16, 2004, Paper P5.1, CD-ROM.

Dansgaard, W., 1964, Stable isotopes in precipitation: Tellus, v. 16, p. 436–469.

Davis, S.N., Whittemore, D.O., Fahryka-Martin. J., 1998, Uses of chloride-bromide ratios in studies of potable water: Ground Water, v. 36, p. 338–350.

Dibblee, T.W., Jr., 1981a, Geologic map of the Palm Springs (15 minute) quadrangle, California: South Coast Society Geologic Map SCGS-3, scale 1:62,500.

Dibblee, T.W., Jr., 1981b, Geologic map of the Idyllwild (15 minute) quadrangle, California: South Coast Society Geologic Map SCGS-5, scale 1:62,500.

Dibblee, T.W., Jr., 2004, Geologic map of the Palm Springs quadrangle, Riverside County, California: Dibblee Geology Center Map #DF-123, scale 1:24,000.

Dobrin, M.B., and. Savit, C.H., 1988, Introduction to geophysical prospecting, McGraw-Hill Book Co., San Francisco, California, 867 p.

Donahue, D.J., Linick, T.W., and Jull, A.J.T., 1990, Ratio and background corrections for accelerator mass spectrometry radiocarbon measurements: Radiocarbon, v. 32, p. 135–142.

Dutcher, L.C., 1953, Memorandum on the flow of Agua Caliente spring after road improvement at Palm Springs, California: U.S. Geological Survey Open-File Report, 7 p.

Dutcher, L.C. and Bader, J.S., 1963, Geology and hydrology of Agua Caliente Spring, Palm Springs, California: U.S. Geological Survey Water-Supply Paper 1605, 43 p.

Eaton, G.F., Hudson, G.B., and Moran, J.E., 2004, Tritium-helium-3 age-dating of groundwater in the Livermore Valley of California: American Chemical Society ACS Symposium Series no. 868, p. 235–245.

Epstein, S. and Mayeda, T., 1953, Variation of O-18 content of water from natural sources: Geochimica Cosmochimica Acta, v.4, p. 213–224.

Faires, L.M., 1993, Methods of analysis by the U.S. Geological Survey National Water Quality Laboratory—Determination of metals in water by inductively coupled plasma-mass spectrometry: U.S. Geological Survey Open-File Report 92–634, 28 p.

Fishman, M.J., 1993, Methods of analysis by the U.S. Geological Survey National Water Quality Laboratory—Determination of inorganic and organic constituents in water and fluvial sediments: U.S. Geological Survey Open-File Report 93–125, 217 p.

Fishman, M.J., and Friedman, L.C., 1989, Methods for determination of inorganic substances in water and fluvial sediments: U.S. Geological Survey Techniques of Water-Resources Investigations, book 5, chapter A1, 545 p.

Flint, L.E., and Flint, A.L., 2007, Regional analysis of ground-water recharge, in Stonestrom, D.A., Constantz, J., Ferré, T.P.A., and Leake, S.A., ed(s)., Ground-water recharge in the arid and semiarid southwestern United States: U.S. Geological Survey Professional Paper 1703, p. 29–60.

Freeze, R.A., and Cherry, J.A., 1979, Groundwater: Englewood Cliffs, N.J., Prentice-Hall, 604 p.

Friedman, I., and O'Neil, J.R., 1977, Chapter KK. Compilation of stable isotope fractionation factors of geochemical interest, in Fleischer, M., ed., Data of Geochemistry (6th ed.): U.S. Geological Survey Professional Paper, 440-KK, 12 p., 49 fig.

Friedman, I., Smith, G.I., Gleason, J.D., Warden, A., and Harris, J.M., 1992, Stable isotope composition of waters in southeastern California 1. Modern precipitation: Journal of Geophysical Research, v. 97, Issue D5, p. 5795–5812.

Galloway, D.L., Jones, D.R., and Ingebritsen, S.E., 1999, Land subsidence in the United States: U.S. Geological Survey Circular 1182, 175 p.

Galloway, D.L., Jones, D.R., and Ingebritsen, S.E., 2000, Measuring land subsidence from space: U.S. Geological Survey Fact Sheet 051–00, 4 p.

Garbarino, J.R., 1999, Methods of analysis by the U.S. Geological Survey National Water Quality Laboratory—determination of dissolved arsenic, boron, lithium, selenium, strontium, thallium, and vanadium using inductively coupled plasma-mass spectrometry: U.S. Geological Survey Open-File Report 99–093, 31 p.

Garbarino, J.R., Kanagy, L.K., and Cree, M.E., 2006, Determination of elements in natural-water, biota, sediment and soil samples using collision/reaction cell inductively coupled plasma-mass spectrometry: U.S. Geological Survey Techniques and Methods, book 5, sec. B, chap. 1, 88 p.

Garrett, A.A., and Dutcher, L.C., 1951, Possible effect of a road improvement on the flow of the Agua Caliente Spring, at Palm Springs, Riverside County, California: U.S. Geological Survey, Ground Water Branch, 8 p.

Gleason, J.D., Veronda, G., Smith, G.I., Friedman, I., and Martin, P., 1994, Deuterium content of water from wells and perennial springs, southeastern California: U.S. Geological Survey Hydrologic Investigations Atlas HA-727.

Godwin, H., 1956, Half-life of radiocarbon: Nature, v. 195, p. 984.

Gonfiantini, R., 1978, Standards for stable isotope measurements in natural compounds: Nature, v. 271, p. 534–536.

Halford, K.J., and Kuniansky, E.L., 2002, Documentation of spreadsheets for the analysis of aquifer-test and slug-test data: U.S. Geological Survey Open-File Report 02–197, 51 p.

Heaton, T.H.E., and Vogel, J.C., 1981, "Excess air" in groundwater: Journal of Hydrology, v. 50, p. 201–216.

Hole, J.A., 1992, Nonlinear high-resolution three-dimensional seismic traveltime tomography: Journal of Geophysical Research, v. 97, p. 6553–6562.

Ingraham, N.L., and Taylor, B.E., 1991, Light stable isotope systematics of large-scale hydrologic regimes in California and Nevada: Water Resources Research, v. 27, p. 77–90.

International Union of Geodesy and Geophysics, 1971, Geodetic Reference System 1967: International Association of Geodesy Special Publication no. 3, 116 p.

Izbicki, J.A., Borchers, J.W., Leighton, D.A., Kulongoski, J., Fields, L., Galloway, D.L., and Michel, R.L., 2003, Hydrogeology and geochemisty of aquifers underlying the San Lorenzo and San Leandro area of the East Bay Plain, Alameda County, California: U.S. Geological Survey Water-Resources Investigations Report 02–4259, 86 p.

Jachens, R.C., and Moring, B.C., 1990, Maps of the thickness of Cenozoic deposits and the isostatic residual gravity over basement for Nevada: U.S. Geological Survey Open-File Report 90–404, 15 p., 2 pl.

Jennings, C.W., 1977, Geologic Map of California: California Division of Mines and Geology Geologic Data Map No. 2, scale 1:750,000

Jennings, C.W., 1994, Fault activity map of California and adjacent areas, with locations and ages of recent volcanic eruptions: California Department of Conservation, Division of Mines and Geology Geologic Data Map No, 6, scale 1:750,000.

Jull, A. J.T., Burr, G.S., McHargue, L.R., Lange, T.E., Lifton, N.A., Beck, J.W., Donahue, D., and Lal, D., 2004, New frontiers in dating of geological, paleoclimatic and anthropological applications using accelerator mass spectrometric measurements of ^{14}C and ^{10}Be in diverse samples: Global & Planetary Change, v. 41, p. 309–323.

Kharaka, Y.K., Gunter, W.D., Aggarwal, P.K., Perkins, E.H., and DeBraal, J.D., 1988, SOLMINEQ.88: a computer program for geochemical modeling of water-rock interaction: U.S. Geological Survey Water-Resources Investigations Report 88–4227, 170 p.

Kharaka, Y.K., and Mariner, R.H., 1989, Chemical geothermometers and their application to formation waters in sedimentary basins, in Naeser, N.D., and McCulloh, T., ed(s)., Thermal History of Sedimentary Basins: Springer Verlag, New York, p. 99–117.

Land, M., Reichard, E.G., Crawford, S.M., Everett, R.R., Newhouse, M.W., and Williams, C.F., 2004, Groundwater quality of coastal aquifer systems in the West Coast Basin, Los Angeles County, California, 1999–2002: U.S. Geological Survey Scientific Investigations Report 2004–5067, 80 p.

Langenheim, V.E., Jachens, R.C., Matti, J.C., Hauksson, E. Morton, D.M., and Christensen, Allen, 2005, Geophysical evidence for wedging in the San Gorgonio Pass structural knot, southern San Andreas fault zone, southern California: Geological Society of America Bulletin, v. 117, no. 11, p. 1554–1572.

Leivas, E., Martin, R.C., Higgins, C.T., and Bezore, S.P., 1981, Reconnaissance geothermal resource assessment of 40 sites in California: California Department of Conservation, Division of Mines and Geology Open-File Report 82–4 SAC, 243 p.

Loeltz, O.J., Ireland, B., Robinson, J.H., and Olmsted, F.H., 1975, Geohydrologic reconnaissance of the Imperial Valley, California: U.S. Geological Survey Professional Paper 486-K, 54 p.

Londquist, C.J., and Martin, Peter, 1991, Geohydrology and ground-water-flow simulation of the Surprise Spring Basin Aquifer System, San Bernardino County, California: U.S. Geological Survey Water-Resources Investigations Report 89–4099, 41 p.

Lucas, L.L. and Unterweger, M.P., 2000, Comprehensive review and critical evaluation of the half-life of tritium: Journal of Research of the National Institute of Standards and Technology, v. 105, p. 541–549.

Mamyrin, B.A., Anufriyev, G.S., Kamenskii, I.L., and Tolstikhin, I.N., 1970, Determination of the isotopic composition of atmospheric helium: Geochemistry International, v. 7, p. 498–505.

Manning, A.H. and Solomon, D.K., 2003, Using noble gases to investigate mountain-front recharge: Journal of Hydrology, v. 275, p. 194–207.

McCleskey, R.B., Nordstrom, D.K., and Ball, J.W., 2003, Metal interferences and their removal prior to the determination of As(T) and As(III) in acid mine waters by hydride generation atomic absorption spectrometry: U.S. Geological Survey Water-Resources Investigations Report 03–4117, 14 p.

McLain, B.J., 1993, Methods of analysis by the U.S. Geological Survey National Water Quality Laboratory; determination of chromium in water by graphite furnace atomic absorption spectrophotometry: U.S. Geological Survey Open-File Report 93–449, 16 p.

Morelli, C., 1974, The International Gravity Standardization Net, 1971: International Association of Geodesy Special Publication no. 4, 194 p.

Morton, R.A., Leach, M.P., Paine, J.G., and Cardoza, M.A, 1993, Monitoring beach changes using GPS surveying techniques: Journal of Coastal Research, v. 9, no. 3, p. 707–720.

Moyle, W.R., Jr., 1974, Temperature and chemical data for selected thermal wells and springs in southeastern California: U.S. Geological Survey Water-Resources Investigations Report 73–33, 31 p.

Nalder, I.A., and Wein, R.W., 1998, Spatial interpolation of climatic normals: test of a new method in the Canadian boreal forest: Agriculture and Forest Meteorology, v. 4, p. 211–225.

Ostlund, H.G. and Werner, E., 1962, The electrolytic enrichment of tritium and deuterium for natural tritium measurements, in Tritium in the Physical and Biological Sciences, v. 1: International Atomic Energy Agency, Vienna, p. 95–104.

Parkhurst, D.L. and Appelo, C.A.J., 1999, User's guide to PHREEQC (Version 2)—A computer program for speciation, batch-reaction, one-dimensional transport, and inverse geochemical calculations: U.S. Geological Survey Water-Resources Investigations Report 99–4259, 310 p.

Patton, C.J. and Kryskalla, J.R., 2003, Methods of analysis by the U.S. Geological Survey National Water Quality Laboratory—Evaluation of alkaline persulfate digestion as an alternative to Kjeldahl digestion for determination of total and dissolved nitrogen and phosphorus in water: U.S. Geological Survey Water-Resources Investigations Report 03–4174, 33 p.

Patton, C.J., and Truitt, E.P., 1992, Methods of analysis by the U.S. Geological Survey National Water Quality Laboratory—Determination of total phosphorus by a Kjeldahl digestion method and an automated colorimetric finish that includes dialysis: U.S. Geological Survey Open-File Report 92–146, 39 p.

Patton, C.J., and Truitt, E.P., 2000, Methods of analysis by the U.S. Geological Survey National Water Quality Laboratory—Determination of ammonium plus organic nitrogen by a Kjeldahl digestion method and an automated photometric finish that includes digest cleanup by gas diffusion: U.S. Geological Survey Open-File Report 00–170, 31 p.

Piper, A.M., 1944, A graphic procedure in the geochemical interpretation of water analyses: American Geophysical Union Transactions, v. 25, p. 914–923.

Plouff, Donald, 1977, Preliminary documentation for a FORTRAN program to compute gravity terrain corrections based on topography digitized on a geographic grid: U.S. Geological Survey Open-File Report 77–535, 45 p.

Poland, J.F., and Dutcher L.C., 1953, Second Memorandum on the flow of Agua Caliente Spring after road construction at Palm Springs, California: U.S. Geological Survey CAL-53-3, 8 p.

Porcelli, D., Ballentine, C.J., and Wieler, R., 2002, An overview of noble gas geochemistry and cosmochemistry, ch. 1, in Porcelli, D., Ballentine, C.J., and Wieler, R., (ed)s, Noble Gases in Geochemistry and Cosmochemistry, Reviews in Mineralogy & Geochemistry, Volume 47: The Mineralogical Society of America, Washington, DC, p. 1–19.

Proctor, R.J., 1968, Geology of the desert hot springs–upper Coachella Valley area, California: California Division of Mines and Geology, Special Report 94, 50 p.

Pruess, K., Oldenburg, C., and Moridis, G., 1999, TOUGH2 User's Guide, Version 2.0: Report LBNL-43134, Lawrence Berkeley National Laboratory, Berkeley, Calif.

Rantz, S.E. and others, 1982, Measurement and computation of streamflow: v. 1, Measurement of stage and discharge, ch.. 8. Measurement of discharge by miscellaneous methods: U.S. Geological Survey Water Supply Paper 2175, p. 260–272.

Reichard, E.G. and Meadows J.K., 1992, Evaluation of a ground-water flow and transport model of the Upper Coachella Valley, California: U.S. Geological Survey, Water Resources Investigations Report 91–4142, 101 p.

Rogers, T.H., 1966, Geologic map of the Santa Ana quadrangle, California: California Division of Mines and Geology Geologic Map, scale 1:250,000.

Rose, T.P., Davisson, M.L., and Criss, R.E., 1996, Isotope hydrology of voluminous cold springs in fractured rock from an active volcanic region, northeastern California: Journal of Hydrology, v. 179, p. 207–236.

Schneider, W.A., 1978, Integral formulation for migration in 2 and 3 dimensions: Geophysics, 43, p. 49–76.

Schon, J. H., 1996, Physical properties of rocks: Fundamentals and principals of petrophysics: Handbook of Geophysical Exploration, Seismic Exploration, v. 18, Elsevier Science, Tarrytown, New York, 600 p.

Schroeder, R.A., Anders, R., Böhlke, J.K., Michel, R.L., and Metge, D.W., 1997, Water quality at production wells near artificial-recharge basins in Montebello Forebay, Los Angeles County, in Kendall, D.W., (ed), Proceedings of the AWRA Symposium, Conjunctive Use of Water Resources: Aquifer Storage and Recovery, American Water Resources Association 33rd National Meeting, Long Beach, California, October 19–23, 1997: American Water Resources Association, Herndon, Virginia, TPS-97-2, p. 273–284.

Schroeder, R.A., Orem, W.H., and Kharaka, Y.K., 2002, Chemical evolution of the Salton Sea, California: nutrient and selenium dynamics: Hydrobiologia, v. 473, p. 23–45.

Smith, G.I., Friedman, I., Gleason, J.D., and Warden, A., 1992, Stable isotope composition of waters in southeastern California: 2. groundwaters and their relation to modern precipitation: Journal of Geophysical Research, v. 97, p. 5813–5823.

Sneed, Michelle, and Brandt, J.T., 2007, Detection and measurement of land subsidence using Global Positioning System and Interferometric Synthetic Aperture Radar, Coachella Valley, California, 1996-2005: U.S. Geological Survey Scientific Investigations Report 2007–5251. Available at http://pubs.usgs.gov/sir/2007/5251/.

Struzeski, T.M., DeGiacomo, W.J., and Zayhowski, E.J., 1996, Methods of analysis by the U.S. Geological Survey National Water Quality Laboratory—Determination of dissolved aluminum and boron in water by inductively coupled plasma-atomic emission spectrometry: U.S. Geological Survey Open-File Report 96–149, 17 p.

Taylor, C.B., and Roether, W., 1982, A uniform scale for reporting low-level tritium in water, in Methods of low-level counting and spectrometry: International Atomic Energy Agency, Vienna, p. 303–323.

Telford, W.M., Geldart, L.O., Sheriff, R.E., and Keyes, D.A., 1976, Applied Geophysics: New York, Cambridge University Press, 960 p.

Thatcher, L.L., Janzer, V.J., and Edwards, K.W., 1977, Methods for the determination of radioactive substances in water: U.S. Geological Survey Techniques of Water-Resources Investigations, chap. A5, 95 p.

Timme, P.J., 1995, National Water Quality Laboratory 1995 services catalog: U.S. Geological Survey Open-File Report 95–352, 120 p.

Turekian, K.K., and Wedepohl, K.H., 1961, Distribution of the elements in some major units of the Earth's crust: Geological Society of America Bulletin, v. 72, p.175–182.

U.S. Geological Survey, 2006, National field manual for the collection of water-quality data: U.S. Geological Survey Techniques of Water-Resources Investigations, book 9, chaps. A1–A9, accessed on July 7, 2006, at http://pubs.water.usgs.gov/twri9A/.

Waring, G.A., 1915, Springs of California: U.S. Geological Survey Water-Supply Paper 338, 410 p.

Weiss, R.F., 1968, Piggyback sampler for dissolved gas studies on sealed water samples: Deep Sea Research, v. 15. p. 721–735.